Modelling Covariances and Latent Variables using EQS

Modelling Covariances and Latent Variables using EQS

G. Dunn, B. Everitt and A. Pickles
Biostatistics Department and MRC Child Psychiatry Department
Institute of Psychiatry
London, UK

CHAPMAN & HALL
London · Glasgow · New York · Tokyo · Melbourne · Madras

Published by Chapman & Hall, 2–6 Boundary Row, London SE1 8HN

Chapman & Hall, 2–6 Boundary Row, London SE1 8HN, UK

Blackie Academic & Professional, Wester Cleddens Road, Bishopbriggs, Glasgow G64 2NZ, UK

Chapman & Hall Inc., 29 West 35th Street, New York NY10001, USA

Chapman & Hall Japan, Thomson Publishing Japan, Hirakawacho Nemoto Building, 6F, 1-7-11 Hirakawa-cho, Chiyoda-ku, Tokyo 102, Japan

Chapman & Hall Australia, Thomas Nelson Australia, 102 Dodds Street, South Melbourne, Victoria 3205, Australia

Chapman & Hall India, R. Seshadri, 32 Second Main Road, CIT East, Madras 600 035, India

First edition 1993

Typeset in 11/13 pt Times by Best-set Typesetter Ltd., Hong Kong
Printed in Great Britain by T. J. Press (Padstow) Ltd., Padstow, Cornwall

ISBN 0 412 48990 2

A catalogue record for this book is available from the British Library

Library of Congress Cataloging-in-Publication data available

Printed on acid-free text paper, manufactured in accordance with ANSI/NISO Z39.48-1992 and ANSI/NISO Z39.48-1984 (Permanence of Paper).

Contents

Foreword

It is a real pleasure to introduce this wonderfully readable approach to structural equation modeling. As you will see, you are in for a treat. No doubt you know that structural modeling has become one of the most popular multivariate statistical methodologies for data analysis. Not only are there few alternatives to this methodology for testing theories with nonexperimental data, but modeling also provides extremely effective tools for data exploration. When you complete this text, you will be able to specify your theory in modeling form and evaluate its appropriateness on real data. Although practical issues in the modeling process represent the main focus of this text, some attention also is given to theoretical topics.

Without scaring you away, I would like to provide my own version of the abstract concepts involved in modeling, and show how these translate into the topics addressed in this text. The basic idea is very general and quite abstract. Suppose that the distribution of variables x in some population is characterized by a vector $\boldsymbol{\sigma}$ of parameters. The structural modeling hypothesis is that this vector of parameters is a function of a smaller number of more basic parameters given in the vector $\boldsymbol{\theta}$, that is, $\boldsymbol{\sigma} = \boldsymbol{\sigma}(\boldsymbol{\theta})$. The task of modeling is to estimate the parameters $\boldsymbol{\theta}$, to evaluate the null hypothesis, and to study alternative hypotheses with appropriate sample data. An obvious extension takes the approach to modeling several populations.

In practice, you already will have done structural modeling in other contexts, but you may not know this. Structural modeling is done under many guises, in many statistical contexts, and usually without a recognition that a generic statistical problem is being addressed. For example, a test of the equality of two means is concerned with the vector $\boldsymbol{\mu}' = (\mu_1, \mu_2)$ and the null hypothesis that $\boldsymbol{\mu} = \boldsymbol{\mu}(\boldsymbol{\theta})$, where $\boldsymbol{\theta} = \mu_1 = \mu_2$. Log-linear structural models have become commonplace in the analysis of categorical data.

The key parameters of the distribution of continuous multivariate variables x are the vector of means μ and the covariance matrix Σ. With multivariate normally distributed variables, these are the only parameters of interest because they contain all the statistical information. This can also be considered to be the case with ordered polytomous variables, when these are considered to be based on underlying continuous multivariate normal variables that have been categorized to yield the observed responses. In more general distributions, the means and covariances also are fundamentally important, although additional information may be available in higher-order moments. This additional information is only rarely studied.

If μ and Σ are the key parameters of many distributions of interest, the structural modeling perspective immediately focuses attention on possible underlying parameters $\boldsymbol{\theta}$ that might generate μ and Σ. Often there is no interest in the means, so that the sample mean is taken as the estimator of μ, and the structural hypothesis focuses on the covariance structure model $\Sigma = \Sigma(\boldsymbol{\theta})$. In mean and covariance structure models, both means and covariances are structured in terms of the basic parameters: $\mu = \mu(\boldsymbol{\theta})$ and $\Sigma = \Sigma(\boldsymbol{\theta})$. In practice, standard statistical criteria are used to estimate $\hat{\boldsymbol{\theta}}$, and the null hypothesis is accepted if $\hat{\mu} = \mu(\hat{\boldsymbol{\theta}})$ and $\hat{\Sigma} = \Sigma(\hat{\boldsymbol{\theta}})$ are close to the sample means and covariances.

In practice, of course, the modeling hypotheses $\mu = \mu(\boldsymbol{\theta})$ and $\Sigma = \Sigma(\boldsymbol{\theta})$ are too abstract for data analysis, since the functional form relating $\boldsymbol{\theta}$ to μ and Σ could be anything. A more restricted but more useful class of models is obtained by considering only those models that result from the specification of a series of restricted linear equations to relate hypothetical latent and observed variate scores. These 'structural equations' imply a structure for μ and Σ. A popular model that combines psychometric factor analysis and econometric simultaneous equation models was developed in the early 1970s by Keesling and Wiley. It is implemented in Jöreskog and Sörbom's LISREL program. However, with its eight or more parameter matrices that yield elements of the vector $\boldsymbol{\theta}$, the model can be overwhelmingly complex, as well as unnecessarily restricted. A simpler yet more general approach was developed by myself with Weeks in the late 1970s. This approach considered the idea that there is a 1:1 correspondence between models that can be specified in path diagram form and those that can be represented by linear equations relating variables. In this approach (Bentler, P. M., and Weeks, D. G. (1980) Linear structural equations with latent variables. *Psychometrika*, **45** 289–308), *the parameters of any linear structural model are the regression coefficients in the linear equations, and the variances and covariances of independent variables*. So, in any covariance structure model, $\boldsymbol{\theta}$ contains only regression coefficients and variances and covariances. In models with mean structures, the parameters in $\boldsymbol{\theta}$ also include the *means of the independent variables and the intercepts of the dependent variables*. As you learn to understand this fundamental idea, you also will understand the basic theory that lies behind the EQS program. The theory is described in mathematical detail in Chapter 10 of the EQS **Manual**, but the technical aspects are not needed to set up and evaluate models in practice.

The original EQS program (Bentler, P. M. (1992) **EQS Structural Equations Program Manual**. Cork, Ireland: BMDP Statistical Software) and its graphical and categorical-variable extension (Bentler, P. M., and Wu, E. J. C. (1993) **EQS/Windows User's Guide, Version 4**. Cork, Ireland: BMDP Statistical Software) were developed to implement the Bentler–Weeks theory in an especially user-friendly way. You should be sure to learn how to use path diagrams to accurately specify your theory. Then, if you have a basic acquaintance with regression equations and regression coefficients, as well as with variances and covariances, with a little help from this book you should be able to set up, run, and evaluate any structural modeling theory, no matter how complex.

The authors give you a systematic and thoughtful way to build up your knowledge base on the variety of models that seem to be useful in practice, as well as the specific aspects of methodology that can help you to creatively implement and evaluate your ideas. If at all possible, you should run their various examples on your own computer, perhaps with variants of the models proposed. Have fun!

Peter M. Bentler
Los Angeles

Preface

This book has been designed as a self-instructional text which serves to introduce the reader to both the principles of statistical modelling of covariance matrices and the use of a particular software package, EQS (Bentler, 1989). In preparing the text, we have assumed that the reader already has some familiarity with elementary statistical methods and with the use of computers, but we do include background material in the earlier chapters for those who are not familiar with covariances (and covariance matrices) and with the principles of statistical modelling. In working through the book, readers are encouraged to run all of the EQS jobs themselves, both as an aid to the understanding of the theoretical material and to get practical experience in the use of the software. Beginners will need to start at Chapter 1 and move through each of the chapters in turn. Readers who have some familiarity with topics such as confirmatory factor analysis and structural equation modelling might wish to start with Chapter 2 (to familiarize themselves with EQS), and then move on to the core of the book (Chapters 4–7). Chapter 8 discusses several practical problems which arise in attempts to analyse *real* sets of data, and suggests useful approaches to their solution.

Acknowledgement and note

We would like to thank Peter Kay for carefully checking all the examples.

The statistical modelling program EQS was developed by Professor Peter M. Bentler and is marketed by BMDP Statistical Software, Inc. Enquiries concerning the EQS program should be addressed to:

BMDP Statistical Software, Inc.
1440 Sepulveda Boulevard
Suite 316
Los Angeles, CA 90025

or

BMDP Statistical Software
Cork Technology Park
Model Farm Road
Cork, Ireland

Overview

Essentially, the text comprises three distinct parts. The first three chapters introduce the reader to the basic ideas and language of covariance structure modelling. They also introduce the EQS software package. Chapters 4–7 make up the second section and cover a wide variety of models suitable for cross-sectional and longitudinal data. They include models for the simultaneous analysis of means and covariances as well as the more familiar models for covariances (correlations) alone. They also describe the simultaneous fitting of models to covariance matrices arising from two or more samples or groups of subjects. Finally, forming a section on its own, Chapter 8 discusses a wide variety of practical problems and suggestions for their solution, including statistical power and sample size estimation, missing data values and non-normality of measurements.

The content of each chapter will now be briefly reviewed to provide the reader with an overall picture of the text.

CHAPTER 1. PRELUDE: THE IDEAS OF COVARIANCE AND OF COVARIANCE STRUCTURE

The book starts with a leisurely introduction to the concept of covariance and to the methods of estimation of covariances. Variances and correlations are then introduced as special cases of covariance. The derivation of expected or predicted values of variances and covariances is then illustrated by a simple statistical model to describe the calibration of guesses of the lengths of pieces of string on ruler measurements. The problems of bivariate calibration when both variables are subject to error are then introduced to illustrate the concepts of **identification** and **under-identification**.

Finally, we discuss elementary examples of **factor analysis** models which are used to explain patterns of covariance between three or more fallible measurements. Each statistical model is illustrated by means of a **path diagram**. Throughout the chapter, the distinction is drawn between **manifest** (observed) and **latent** (not directly observable) variables or measurements, and also, in the context of the statistical models, between **dependent** variables and **independent** variables.

CHAPTER 2. WRITING A SIMPLE EQS PROGRAM

Here we slowly work through all of the statements and commands required to set up an EQS input file. They are discussed one at a time and then put together to make a completed program. After working through this chapter the reader should have an understanding of the basic vocabulary and syntax of EQS, and should also be able to write and execute simple EQS jobs. The structure and interpretation of EQS output files are also explained in detail.

CHAPTER 3. STATISTICAL MODELLING IN EQS

Having covered the basic language of covariance modelling and of the EQS software, the discussion then moves on to cover general strategies for model fitting and hypothesis testing. Two fundamental types of model are discussed: **confirmatory factor models** and **structural equation models**. Fitting criteria, goodness-of-fit statistics, and methods of imposing constraints on parameter estimates are all discussed with reference to real data sets and associated EQS runs. At the end of the chapter we explain how EQS reacts to under-identification and to impermissible parameter estimates. Finally, models involving the simultaneous analysis of means and covariances are introduced and illustrated by estimating the intercept parameter in a simple bivariate regression problem.

CHAPTER 4. CONFIRMATORY FACTOR ANALYSIS MODELS

In this chapter a more complex example of a confirmatory factor analysis model is described in detail. Problems with a poorly specified model are encountered and statistical methods for suggesting modified models are discussed. The use of correlations between parameter estimates for indicating problematical models is illustrated.

CHAPTER 5. MULTITRAIT-MULTIMETHOD AND MULTIPLE INDICATOR MULTIPLE CAUSE MODELS

Two types of model are discussed in this chapter. The first is a special type of confirmatory factor analysis model, used in situations where a number of different methods are used to measure a number of traits of interest. The second involves situations where several observed variables are assumed to 'cause' a latent variable which itself is indicated by a number of other observed variables. A number of problems specific to these types of model are described and illustrated.

CHAPTER 6. MODELS FOR LONGITUDINAL DATA

Longitudinal data are data that involve the repeated measurement of the same set of subjects over time. An enormous variety of research questions can be addressed with such data, ranging from questions about the extent of continuity in development, to the pattern and direction of effects among a set of variables recorded over time. This chapter applies and develops the modelling techniques of the previous chapters to such data, emphasizing two main themes. The first is the extent to which measurement error can obscure very simple underlying processes, making them appear more complicated than they really are. The second is an emphasis on models of the process and their testing, rather than the simple empirical derivation of some 'parsimonious' description of the data. The chapter also introduces the concepts and techniques for estimating **direct** and **indirect effects**, **reciprocal effects** and **missing variables**.

CHAPTER 7. SIMULTANEOUS ANALYSIS OF TWO OR MORE GROUPS

Many applications of covariance modelling procedures occur in situations where the data fall into a number of natural groups.

Examples are gender, diagnostic category, race and marital status. In such situations, it is often of interest to compare models across groups to determine which parameters are equal and which are different. Here a number of multigroup examples are described. In addition, an explanation of how to assess differences in means of latent variables in the different groups is given.

CHAPTER 8. COMMON PRACTICAL PROBLEMS

There is often a gulf between being able to fit models to the neat and straightforward examples of textbooks and successfully tackling one's own data sets. This chapter attempts to fill this gulf by firstly giving some practical hints on model fitting, and secondly describing a number of common practical problems and their possible solutions. These common problems include **non-normal** and **categorical** data, **missing** data, and **complex sample designs**. In addition, the chapter discusses the concept of **statistical power**, and how it may be examined for the kinds of model in this book.

Prelude: the ideas of covariance and of covariance structure

1

1.1 INTRODUCTION

This primer starts by introducing you to the basic language and terminology of covariance structure modelling. It is assumed that you have some intuitive understanding of random variation and of random variables, and we start by discussing methods of measuring the strength of association or relationship between each pair of two or more random variables. The central concept involved in this discussion will be that of **covariance**. Its definition is introduced, together with methods for its estimation from a sample of data. Measures of dispersion or variability (**variances**) are also shown to be a special case of covariance, as are **correlation coefficients**.

In looking at associations between two or more random variables two types of variable can be distinguished. The first can be observed or measured directly and are referred to as **manifest** variables. Those of the second type are **latent** variables. These are characteristics which are not directly observable. They may be straightforward concepts such as height and weight which we explicitly acknowledge cannot be measured without error (that is, an observed measurement is an example of a manifest variable, while the corresponding unknown, but true, value is a latent variable). On the other hand, they may be theoretical concepts that are introduced to explain covariance between the manifest variables or indicator variables. An example of this type is the set of scores on a battery of cognitive tests that are assumed in some way to reflect a subject's cognitive ability or general intelligence. Intelligence is like weight or height in the sense that it is always estimated from fallible measurements but might be thought to be more abstract or distinct from the actual measurements themselves. The distinction, however, is not at all clear-cut and we will often use the more familiar physical examples to motivate discussions of models for analogous behavioural and sociological measurements.

After reading this introductory chapter you should be able write down simple algebraic expressions to describe relationships between two or more random variables. You should also be able to derive expressions describing the expected covariances between these variables. Although we think that the ability to derive expectations for covariances is a useful skill to acquire, we do wish to stress at the outset that it is not an essential one for the use of the EQS progam itself. The program is quite capable of the required calculations. The motivation behind the derivation of expected covariances is to give you insight into what EQS is doing, what are the limitations, and why things might occasionally go wrong.

Many people find that pictures are much more informative than words (or, in this case, algebra). Graphical illustrations are used throughout the text, but, in particular, you will be shown how to draw and interpret **path diagrams**. All models will be described both using simple algebraic expressions and through the use of path diagrams. We will also try to explain what is going on through relatively non-technical English!

1.2 HOW LONG IS A PIECE OF STRING?

Display 1.1 gives information on the lengths of 15 pieces of string. The column headed T provides measurements made with a rule. The columns labeled G, B and A are independent guesses made by Graham, Brian and Andrew, respectively. T, G, B and A are all examples of random variables. The guesses, G, B and A, are clearly examples of error-prone or fallible measures. T, on the other hand, might be regarded as the 'truth' but in reality is also fallible. It is merely much more accurate. All four measurements are, therefore, fallible manifest variables. Furthermore, it is obvious that these four measures will be highly correlated; this correlation arises from the fact that they are all measures of the same concept: length. Corresponding to each piece of string is a

True length (T)	Graham's guess (G)	Brian's guess (B)	Andrew's guess (D)
6.3	5.0	4.8	6.0
4.1	3.2	3.1	3.5
5.1	3.6	3.8	4.5
5.0	4.5	4.1	4.3
5.7	4.0	5.2	5.0
3.3	2.5	2.8	2.6
1.3	1.7	1.4	1.6
5.8	4.8	4.2	5.5
2.8	2.4	2.0	2.1
6.7	5.2	5.3	6.0
1.5	1.2	1.1	1.2
2.1	1.8	1.6	1.8
4.6	3.4	4.1	3.9
7.6	6.0	6.3	6.5
2.5	2.2	1.6	2.0

Measurements for 15 pieces to the nearest tenth of an inch

DISPLAY 1.1 How long is a piece of string?

true but unknown length. Although T will be a more reliable indication of this length than G, B or A, the truth will always remain unknown. The true length, therefore, is a relatively straightforward example of a latent variable. Unlike some of the later behavioural or sociological examples, it does not need much imagination to accept the validity of this particular latent variable!

Now, you may think that this string example is trivial. In some senses it is, but we have deliberately chosen this example to introduce the basic concepts of covariance and covariance structure because of its familiarity and simplicity. In particular, we do not want to put you off covariance structure models because of doubts concerning their validity or utility in understanding a given set of behavioural data. That is a different problem! Too often, potential users of these statistical techniques are put off by seeing them used on poor or inappropriate sets of data. Readers should be careful to distinguish the validity or utility of a particular set of statistical tools from criticisms of a particular example of their use.

Returning to the 15 pieces of string, let us start by looking at

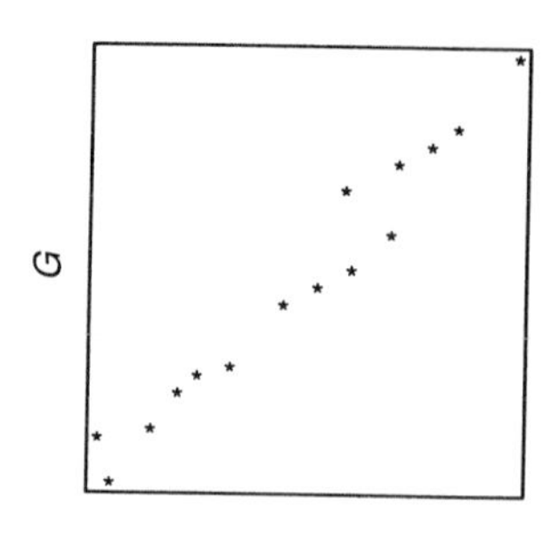

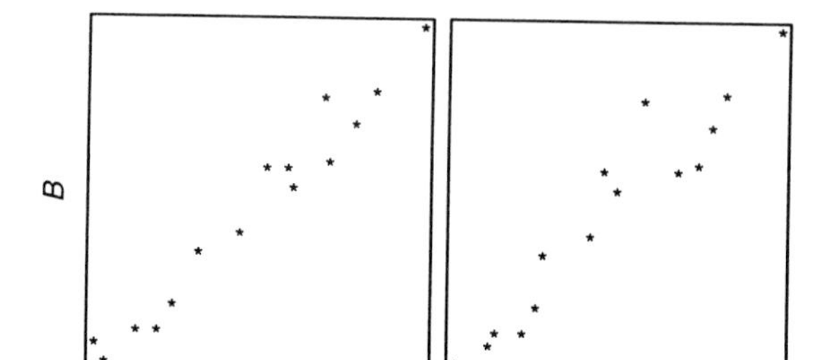

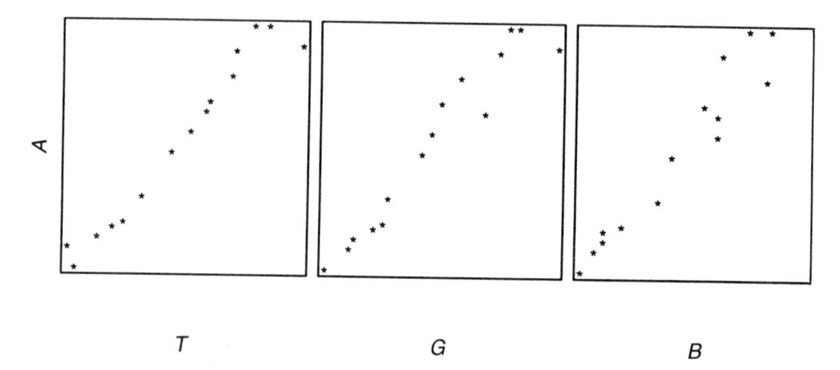

DISPLAY 1.2 Measurements of lengths of string.

scatter diagrams for each possible pair of measurements. These are shown in Display 1.2. It is obvious, again, that all measurements show high covariation. Relatively high values on one are associated with similar high values on another. But how can this covariation be measured? Most readers at this point might suggest the calculation of a correlation coefficient (typically the familiar Pearson product-moment correlation), but here a more general coefficient of covariance will be introduced.

Graham (G)	Brian (B)	$G-\bar{G}$	$B-\bar{B}$	$(G-\bar{G})(B-\bar{B})$
5.0	4.8	1.57	1.37	2.15
3.2	3.1	−0.23	−0.33	0.08
3.6	3.8	0.17	0.37	0.06
4.5	4.1	1.07	0.67	0.72
4.0	5.2	0.57	1.77	1.01
2.5	2.8	−0.93	−0.63	0.59
1.7	1.4	−1.73	−2.03	3.51
4.8	4.2	1.37	0.77	1.05
2.4	2.0	−1.03	−1.43	1.47
5.2	5.3	1.77	1.87	3.31
1.2	1.1	−2.23	−2.33	5.20
1.8	1.6	−1.63	−1.83	2.98
3.4	4.1	−0.03	0.67	−0.02
6.0	6.3	2.57	2.87	7.38
2.2	1.6	−1.23	−1.83	2.25

Cov(*G*,*B*) = mean of the products on the right = 2.12

DISPLAY 1.3 Calculation of covariance of *G* and *B*.

Display 1.3 illustrates the calculation of the covariance between *G* and *B*. First we calculate the sample means for *G* and *B*; these are represented by $\bar{G}$ and $\bar{B}$ respectively. Here $\bar{G}$ and $\bar{B}$ are both equal to 3.43. We then look at the deviations of the guesses for each piece of string from their respective means, that is, $G-\bar{G}$ and $B-\bar{B}$. Display 1.4 provides a plot of $G-\bar{G}$ against $B-\bar{B}$. This is identical to the plot of *G* against *B* in Display 1.2, except that the axes have been moved. By saying that relatively high values of *G* are associated with similarly high values of *B* we are simply saying that the product $(G-\bar{G})\ (B-\bar{B})$ is positive. For any particular piece of string either both $G-\bar{G}$ and $B-\bar{B}$ are positive and hence their product is positive, or both $G-\bar{G}$ and $B-\bar{B}$ are negative and again their product is positive. These so-called cross-products are summarized by taking their average. In general,

$$\mathrm{Cov}(X,Y) = \Sigma(X-\bar{X})(Y-\bar{Y})/N$$

where Cov(X,Y) is the sample covariance of two random variables, X and Y, Σ indicates the summation of the products of the pairs of deviations over the whole sample of measurements, and N

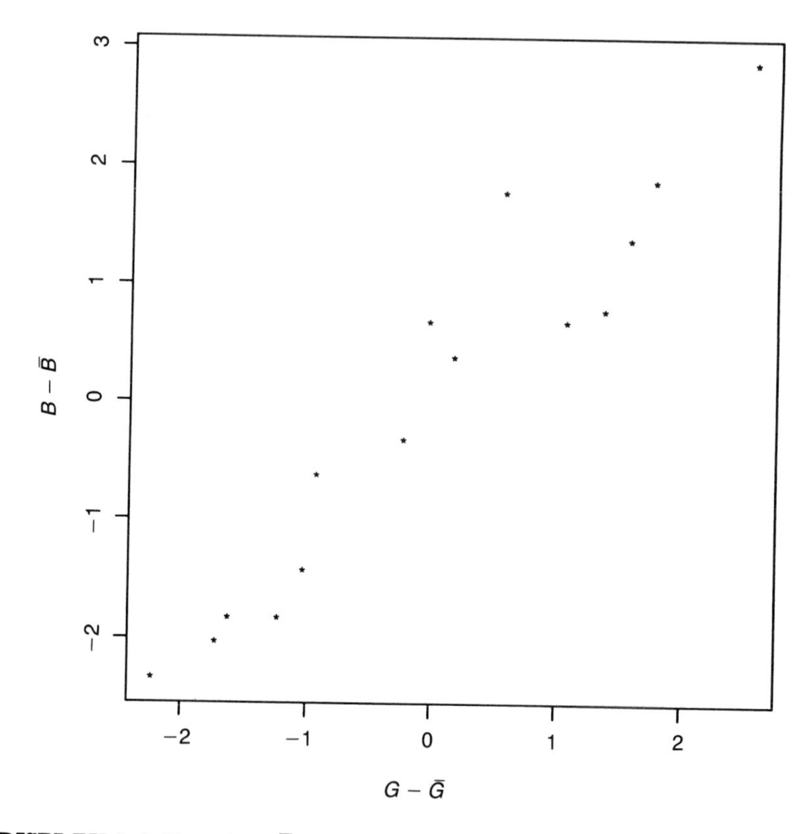

DISPLAY 1.4 Plot of $B-\bar{B}$ against $G-\bar{G}$.

is the size of the sample. For string measurements, G and B, the covariance is about 2.12.

A covariance can have any value ranging from $-\infty$ to $+\infty$. A value of zero indicates that there is no association between two variables (or, more strictly, there is no *linear* association). A positive value is an indication that as one of two variables increases then, on average, so does the other. A negative covariance indicates that as one variable increases the other decreases, and *vice versa*.

An alternative and more widely used, measure of covariance is

$$\text{Cov}(X,Y) = \Sigma(X-\bar{X})(Y-\bar{Y})/(N-1)$$

where the sum of products of the paired deviations is divided by $N-1$ rather than N. This is for relatively subtle technical reasons.

It can be shown that division by N gives a statistic which, when considered as an estimator of the corresponding population parameter, is slightly biased. Division by $N-1$ instead rectifies this bias. Using this covariance formula, the covariance of Graham's and Brian's guesses is about 2.27. Note that if we look at the covariance of a variable, X, with itself, then we obtain either

$$\mathrm{Cov}(X,X) = \Sigma(X-\bar{X})(X-\bar{X})/N$$
$$= \Sigma(X-\bar{X})^2/N$$

or

$$\mathrm{Cov}(X,X) = \Sigma(X-\bar{X})^2/(N-1)$$

which correspond to the two alternative definitions of a sample **variance**. Another point to note is that if we replace the raw measurements, X and Y, by their corresponding standardized

Variable	*Mean*	*Std dev.*		
Ruler	4.2933	1.9692		
Graham	3.4333	1.4563		
Brian	3.4267	1.6294		
Andrew	3.7667	1.7987		
Covariance matrix[1]				
Lower triangular form				
Ruler	3.8778			
Graham	2.8110	2.1210		
Brian	3.1480	2.2669	2.6550	
Andrew	3.5062	2.5690	2.8341	3.2352
Correlation matrix				
Lower triangular form				
Ruler	1.0000			
Graham	0.9802	1.0000		
Brian	0.9811	0.9553	1.0000	
Andrew	0.9899	0.9807	0.9684	1.0000

[1] Using $N-1$ as the divisor.

DISPLAY 1.5 Summary statistics for string data.

values (deviations of the raw scores from their means, measured in standard deviations) or Z-scores, Z_x and Z_y, then it is straightforward to show that

$$\mathrm{Cov}(Z_x, Z_y) = \mathrm{Corr}(X, Y)$$

where $\mathrm{Corr}(X,Y)$ is the familiar **correlation** between a pair of variables, X and Y. As long as we are consistent, it makes little difference whether N or $N-1$ is used in the denominator. We will usually divide by $N-1$. Display 1.5 gives us summary statistics for our four measures of string length (using the divisor $N-1$ for estimates of both variance and covariance). These are the statistics that are used in covariance structure modelling. Before we can ask how, however, we need to learn a few basic rules for the manipulation of covariances.

Exercise 1.1

Using a similar table to that shown in Display 1.5, calculate the covariance of T and A. Calculate the correlation between T and A directly and then, by first standardizing T and A, show that the correlation is equivalent to $\mathrm{Cov}(T,A)$. Demonstrate that

$$\mathrm{Corr}(T,A) = \mathrm{Cov}(T,A)/\mathrm{sd}(T)\mathrm{sd}(A)$$

where $\mathrm{sd}(T)$ and $\mathrm{sd}(A)$ are the sample standard deviations of T and A, respectively.

Exercise 1.2

Note that the string measurements given in Display 1.1 are in inches. Convert T and A into centimetres (1 inch = 2.54 centimetres) and repeat Exercise 1.1. Comment on your findings.

1.3 SIMPLE MEASUREMENT MODELS: CALIBRATION

Returning to Display 1.1, now consider the calibration of Graham's guesses (G) on what, for the time being, will be regarded as the truth (T). This is done by fitting a simple regression model of the form:

$$G = a + bT + E$$

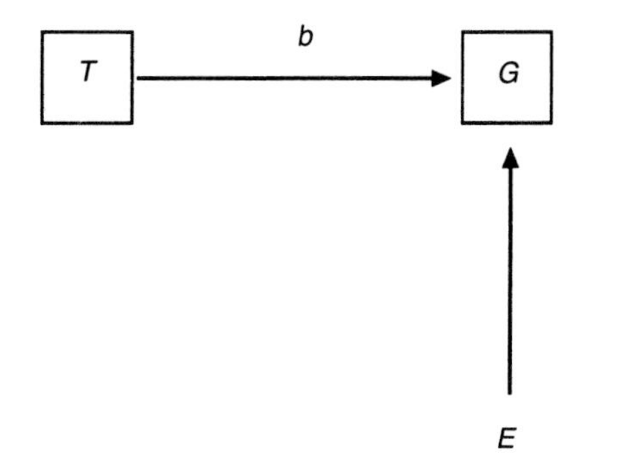

Observed variables, G and T, are both enclosed in square boxes. The measurement error, E, is not. Causal links are represented by single-headed arrows. The regression coefficient is represented by b.

DISPLAY 1.6 Path diagram for a simple bivariate regression model calibrating Graham's guess of string length on the truth (ruler-made measurement).

where a and b are constant coefficients (parameters) and E is a new random variable representing the unsystematic or random error in Graham's guesses. The parameters a and b correspond to the intercept and slope coefficients measuring the systematic bias in Graham's guesses. (If there were no biases relative to T, a would be zero and b would be unity). In this model G is referred to as the **dependent** variable and T is the **independent** variable. The model is illustrated graphically in Display 1.6. In this **path diagram**, the observed variables are enclosed in square boxes and causal links are represented by single-headed arrows. The measurement error, E, is not enclosed by a box. We have assumed that T and E are uncorrelated. It is also assumed that the Es have a mean of zero and a constant variance, Var(E).

Now, this model implies certain predictions for the variance of G and for the covariance of G and T. These are shown in Display 1.7. The aim of the analysis is to compare these expectations with the observed values in order to estimate the values of a and b. Var(E) is also an interesting parameter to estimate. It reflects the precision of Graham's guesses (lower variance corresponding to higher precision).

$$\text{Var}(G) = \text{Var}(a + bT + E)$$
$$\text{Cov}(G,T) = \text{Cov}(a + bT + E,T)$$

DISPLAY 1.7 Expected values for a simple linear regression model.

Consider three random variables, X, Y and Z, and two constants, a and b.

The terms Mean(.), Var(.) and Cov(.) are respectively used to indicate the mean, variance and covariance of the variable combination(s) within the brackets.

Means:

$$\text{Mean}(a + X) = a + \text{Mean}(X)$$
$$\text{Mean}(bX) = b\text{Mean}(X)$$
$$\text{Mean}(a + bX) = a + b\text{Mean}(X)$$

Variances:

$$\text{Var}(a + X) = \text{Var}(X)$$
$$\text{Var}(aX) = a^2\text{Var}(X)$$
$$\text{Var}(X + Y) = \text{Var}(X) + \text{Var}(Y) + 2\text{Cov}(X,Y)$$
$$\text{Var}(X - Y) = \text{Var}(X) + \text{Var}(Y) - 2\text{Cov}(X,Y)$$
$$\text{Var}(aX + bY) = a^2\text{Var}(X) + b^2\text{Var}(Y) + 2ab\text{Cov}(X,Y)$$

For two uncorrelated variables, X and Y:

$$\text{Var}(X + Y) = \text{Var}(X) + \text{Var}(Y)$$
$$\text{Var}(X - Y) = \text{Var}(X) + \text{Var}(Y)$$

Covariances:

$$\text{Cov}(a,X) = 0$$
$$\text{Cov}(aX,bY) = ab\text{Cov}(X,Y)$$
$$\text{Cov}(X,Y + Z) = \text{Cov}(X,Y) + \text{Cov}(X,Z)$$

DISPLAY 1.8 Rules for calculation of means, variances and covariances of functions of random variables.

$$\text{Mean}(G) = a + b\text{Mean}(T)$$
$$\text{Var}(G) = b^2\text{Var}(T) + \text{Var}(E)$$
$$\begin{aligned}\text{Cov}(G,T) &= \text{Cov}(a,T) + \text{Cov}(bT,T) + \text{Cov}(E,T)\\ &= b\text{Var}(T)\end{aligned}$$

DISPLAY 1.9 Derivation of expected values in simple linear regression.

Before we can do this, however, we need some rules for the derivation of means, variances and covariances of random variables. The rules required for our purposes are provided in Display 1.8. We do not explain how these rules are derived, but are content to let the reader use them as given. If you wish to find out about their derivation see, for example, Appendices 1 and 2 of Dunn (1989). You will also notice that there are redundancies built into Display 1.8. We are not here looking for mathematical elegance but simply ease of use. Returning to the calibration problem, the statistical model implies the relationships (expected values) given in Display 1.9. The three expectations in Display 1.9 can be equated with their observed values (given in Display 1.5) in order to solve for the unknown parameter values. The slope of the calibration line b, for example, is estimated by the ratio $\text{Cov}(G,T)/\text{Var}(T)$, which is 0.725; that for the intercept, a, is provided by $G-bT$, which is 0.320. Finally, $\text{Var}(E)$ is estimated by $\text{Var}(G)-b^2\text{Var}(T)$, which is 0.083. Note that the variance of T is here also regarded as a model parameter which, of course, is estimated simply by its corresponding sample value.

Exercise 1.3

Let two random variables, X and Y, have variances of 2 and 6, respectively. Their correlation is 0.5. Calculate the following:

1. $\text{Var}(6Y)$
2. $\text{Cov}(X,Y)$
3. $\text{Cov}(2X,Y)$
4. $\text{Var}(6+2X+4Y)$
5. $\text{Var}(3X-5Y)$

Exercise 1.4

Interpret your findings in Exercise 1.2 by reference to the rules in Display 1.8.

1.4 SIMPLE MEASUREMENT MODELS: CALIBRATION WITH ERRORS IN BOTH VARIABLES

Now let us assume that we only have the guesses of string length provided by Graham (G) and Brian (B); we do not have access to the measured values T and cannot therefore assume that we have any realistic access to the truth. Consider the following pair of simultaneous equations:

$$G = a + bF + E_1$$

and

$$B = c + dF + E_2$$

Here we have postulated an underlying latent variable, F. As in the last example, we have estimates for the means of G and B, their variances and their covariance. In the social and behavioural sciences measurements are usually, at most, interval and the position of zero is arbitrary. Usually, therefore, we are not interested on the intercept terms, a and c, but with the slope parameters, b and d. The measurement model provided by the two equations above is illustrated by the path diagram in Display 1.10. The parameters to be estimated are b, d, Var(F), Var(E_1) and Var(E_2).

We start by assuming that Cov(E_1,E_2), Cov(F,E_1) and Cov(F,E_2) are all zero. In other words, we assume that measurement errors (E_1 and E_2) are uncorrelated and that they are each uncorrelated with the latent variable, F. This may not be an entirely realistic assumption but, if it is not made, then it proves impossible to estimate any of the other parameters of interest. There are now five parameters to be estimated, but only three statistics providing any relevant information (Var(G), Var(B) and Cov(G,B)). We have a problem! The expected variances of G and B are given in Display 1.11, as is their expected covariance.

There are no unique solutions for the parameter estimates. The estimated value of Var(F), can be halved, and correspondingly

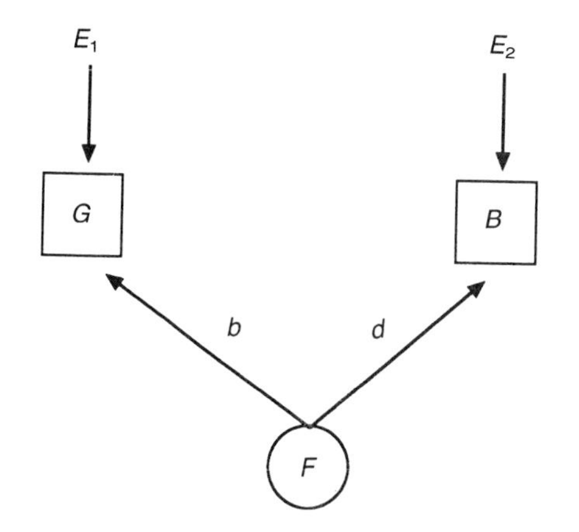

Again causal direction is implied by the single-headed arrows. The underlying latent variable or factor F is enclosed within a circle.

DISPLAY 1.10 Path diagram for a measurement model for two fallible measures of string length (Graham's and Brian's guesses).

$$\mathrm{Var}(G) = b^2\mathrm{Var}(F) + \mathrm{Var}(E_1)$$
$$\mathrm{Var}(B) = d^2\mathrm{Var}(F) + \mathrm{Var}(E_2)$$
$$\mathrm{Cov}(G,B) = bd\mathrm{Var}(F)$$

DISPLAY 1.11 Expected values in regression with both measures subject to error.

both b and d multiplied by $\sqrt{2}$, without altering the predicted variances and covariances of G and B. The model is said to be **under-identified**. It is the classical problem of regression when both variables are subject to error. It can only be solved by the introduction of some sensible constraints on the parameter values. We can start by realizing that G and B can only provide information on the biases of Graham and Brian's guesses *relative to each other*; the ratio of b to d can be estimated but not b and d themselves. This problem can be solved by either arbitrarily fixing b equal to 1 (or, equivalently, d equal to 1) or by fixing the variance of F to be equal to 1 (or any other constant, so fixing the common scale of measurement for G and B). Which of the above constraints is chosen makes no difference to the interpretation of the results. The resulting model, however, is still under-identified. The only way round the problem is additionally either to make b and d equal to each other or to make an assumption concerning the relative sizes of $\text{Var}(E_1)$ and $\text{Var}(E_2)$. The trouble with both of these alternatives is that they involve constraints on aspects of the data which we are specifically interested in investigating! The relative bias of one measure relative to the other should be checked. Similarly, one wishes to test whether the precisions of two measuring instruments are the same. This problem can obviously be rather frustrating. Ideally one should be aware of problems of the identifiability of covariance structure models at the design stage of a research project and modify the design accordingly. One way round the present problem, for example, is to use three independent measures rather than just two. Alternatively, one or both of the two measures could be repeated. The data could then be analysed using a factor analysis model as described below and in the following chapters.

1.5 A SIMPLE FACTOR ANALYSIS MODEL

Now consider a simple measurement model for the three guessed lengths G, B and A. We will again assume that we have no knowledge of the truth. The measurement model is

$$
\begin{aligned}
G &= a + bF + E_1 \\
B &= c + dF + E_2 \\
A &= e + fF + E_3
\end{aligned}
$$

This is illustrated by the path diagram in Display 1.12. Again, ignoring the means and intercept parameters, we have seven parameters to be estimated from the data: the regression coefficients b, d, f and the variances $\text{Var}(F)$, $\text{Var}(E_1)$, $\text{Var}(E_2)$ and $\text{Var}(E_3)$. The six statistics available for use in the parameter estimations are $\text{Var}(G)$, $\text{Var}(B)$, $\text{Var}(A)$, $\text{Cov}(G,B)$, $\text{Cov}(G,A)$ and $\text{Cov}(B,A)$. The model is again under-identified but, if either $\text{Var}(F)$ were constrained to be 1 or one of the regression coefficients constrained to be 1, then the model would be **just identified**. There are now exactly the same number of unconstrained parameters to be estimated as there are informative

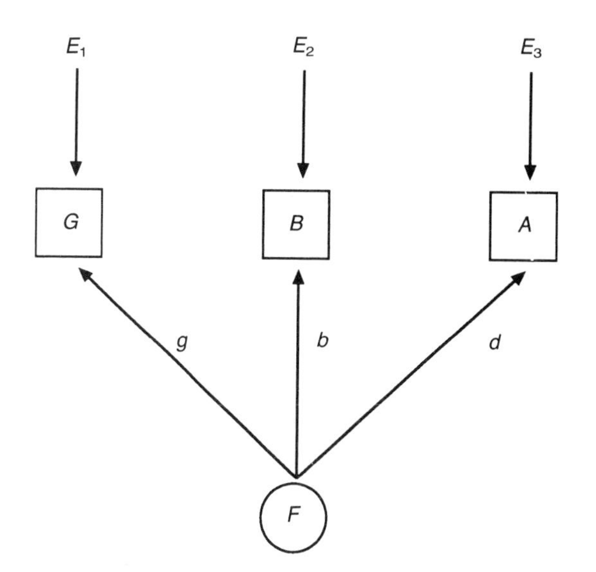

DISPLAY 1.12 Path diagram for a factor analysis model of string lengths.

sample statistics. Equating observed and expected variances and covariances will yield the required estimates.

The above statistical model is an example of Spearman's **common factor model**. The latent independent variable, F, is the single common factor; the random deviations or residuals (E_1, E_2 and E_3) are examples of **specific factors**. In the terminology of factor analysis, the regression coefficients (b, c and d) are **factor loadings**. The error variances ($\text{Var}(E_1)$, $\text{Var}(E_2)$ and $\text{Var}(E_3)$) are known as **specific** variances and the ratio, $b^2\text{Var}(E)/\text{Var}(G)$, for example, is the **communality** (common variance) of Graham's guesses. In this particular example of Spearman's factor analysis model the communality of a particular measure can also be interpreted as its **reliability**. Later chapters will provide more details on the use of factor analysis models.

Exercise 1.5

Estimate the parameters of the above factor model.

Exercise 1.6

Now consider an analogous measurement model involving all four string measures (i.e. including T). Regard T as being fallible in the same way as the others and derive expressions for the expected covariances and covariances. Introduce the constraint Var(F)=1 and proceed to equate expected values with the observed statistics. You should now find that you can obtain *more than one unique estimate* for some of the parameters. The model is said to be **over-identified**. Clearly a better strategy for estimation is needed.

1.6 SUMMARY

We have introduced some simple measurement models for string measurements to illustrate the concept of covariance and ways in which covariances can be used to infer certain characteristics of these measurement models. By now you should be able to write down sets of simple model equations and use these to derive expected variances and covariances. Although they have been used in this preliminary chapter to provide parameter estimates, their main aim is to facilitate exploration of the major problem of under-identification. Over-identification, on the other hand, is not a problem! This provides information (degrees of freedom) by which we can test the fit of the model. These ideas will be developed in later chapters. It is now time to turn to the role of EQS.

Writing a simple EQS program: learning the language

2

2.1 INTRODUCTION

In this chapter we introduce you to the basic and most commonly used commands in EQS. We illustrate these commands by slowly building an input file to fit a simple factor model for the string measurements given in Display 1.1. After reading this chapter you should have an understanding of the basic vocabulary and syntax of EQS and, hopefully, be able to put together your own input files. You should also be able to read and understand input files prepared by other investigators, together with the output produced by running these EQS programs.

2.2 RUNNING EQS

EQS runs as a batch program. This means that in order to fit a model we have to set up an input file containing the appropriate statements and commands. This input file is then read into the EQS package. EQS only then attempts to execute the instructions, sending the results to an output file for later inspection. In this book we will assume that EQS is being run on an IBM-compatible personal computer with MS-DOS. Suppose, for example, that you have created a DOS file containing the EQS instructions. This DOS file is named EXAMPLE.EQS. A program run is started by typing:

EQS ⟨return⟩

You will then receive a prompt from EQS asking you the name of the input file. You then type:

EXAMPLE.EQS ⟨return⟩

Finally, EQS will prompt you for the name of the output file. You then type, for example:

EXAMPLE.OUT ⟨return⟩

As EQS runs this job its progress is monitored on the computer's screen. If the run is successful you will eventually see the statement:

eqs is done

If the run fails for some reason it will be clear from the messages on the screen. Occasionally EQS might get stuck, requiring you to reboot the computer by simultaneously pressing the CTRL, ALT and DEL keys.

EQS generates several sections of output information using a 132-column print format. The package contains a module called BROWSE that enables you to inspect the output (both in the case of a successful run and in case of failure). We will not describe procedures for obtaining hard copy of the output as this will be machine-dependent. To use BROWSE to examine EXAMPLE.OUT we simply type:

BROWSE EXAMPLE.OUT ⟨return⟩

The output appears on the screen and can be examined at leisure. To leave BROWSE simply press ESC.

Readers who have the Windows versions of EQS (to run on an IBM-compatible PC or on an Apple Macintosh, for example) can use the Windows environment for data and file manipulation, preliminary analysis of their data, and to generate EQS input files. The executable EQS program and the corresponding EQS input files are common to all versions of EQS, whether or not the Windows environment is used.

2.3 CREATING AN INPUT FILE

The input file of instructions can be created using an editor or, if preferred, a word-processing package. If a word-processing package is used it must be able to produce a DOS text file which is no wider than 80 columns. Before starting to write your first input file, however, it will be helpful to type the command:

EQSHELP ⟨return⟩

This command initializes the help system, allowing you to get help in writing the input instructions. When help is needed you simply press ALT-H and select from items on the pop-up menus. This is a memory-resident help facility that should be available to you even while you are editing an input file.

The next task is to 'open' or simply start a new file in the editor or your word-processor. The basic structure of the input file is a set of key words, each preceded by a '/' and followed *on the next line*(*s*) by a 'paragraph' of specific information. The 'sentences' of each paragraph are separated by semi-colons(;). Each sentence can extend over more than one line and there may be more than one separate sentence per line. Both upper- and lower-case letters can be used (they are not distinguished in EQS). Any text on a line to

```
/TITLE
 Simple Bivariate Regression
 Graham's guesses against the ruler
 Data from Display 1.1
/SPECIFICATION
 CASES=15;
 ! Number of cases
 VARIABLES=2;
 ! Number of variables
 ME=ML;
 ! Method of estimation
/LABELS
 V1=Graham;
 V2=Ruler;
/EQUATIONS
 V2=0.5*V1 + E2;
 ! Regression equation
/VARIANCES
 E2=1.0*;
 V1=1.0*;
 ! Starting values
/MATRIX
 3.878
 2.811   2.121
 ! The data as a covariance matrix
/END
```

DISPLAY 2.1 A simple EQS input file.

the right of an explanation mark (!) is ignored by EQS and can be used to add explanatory comments to your input file.

An example of a short input file is shown in Display 2.1. You will be left to explain Display 2.1 in a later exercise (see Exercise 2.1). Here, however, we will introduce a basic set of key words and their paragraphs. Others will appear in later chapters when necessary.

2.4 PROGRAM INPUT

As is stated in the EQS manual (Bentler, 1989: p. 44), it is easy to be overwhelmed by the many possibilities in a program such as EQS. There are very many options and the more advanced ones can be a source of confusion, particularly to a beginner. We start, therefore, with a basic set of key words. These are the ones which will be most commonly used, but even these are not always necessary in any given run. Others will be introduced later. The optional keywords are all followed by the word 'Optional' in brackets.

/TITLE (Optional)

This paragraph is identified by the use of the keyword: /TITLE. It is followed by one or more lines of information. An example is shown in Display 2.2. In the output, EQS will print all of the /TITLE paragraph once and will repeat the first line as a header on all pages of the output file. Although /TITLE is optional, it is a good habit for the user to provide information concerning the identity and rationale behind the run for later reference!

/SPECIFICATIONS

In this paragraph the user provides information concerning the number of cases, the number of input variables, and the method (or methods) of estimation required. In addition, you may wish to specify the form of the data (correlation or covariance matrix, for example, or the raw data themselves) and the name of any data file to be read. The paragraph key word may be abbreviated to /SPEC. Display 2.3 gives an example of the use of the

```
/TITLE
Simple Factor Model
Data are guesses of string lengths
From Display 1.1
```

DISPLAY 2.2 An example of the use of the /TITLE paragraph.

```
/SPECIFICATIONS
 CASES=15;
 VARIABLES=4;
 METHOD=ML;
 MATRIX=RAW;
 DATA_FILE=STRING.DAT;
```

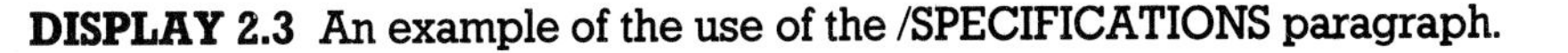

DISPLAY 2.3 An example of the use of the /SPECIFICATIONS paragraph.

/SPECIFICATIONS paragraph. These instructions would be appropriate for reading and analysing the string length data given in Display 1.1. The bottom two statements specify that the raw data can be found in a DOS file called STRING.DAT. The first two instructions complete the data description by telling EQS that there are 15 cases with 4 variables per case. It is assumed that you will be using a free data format (one in which data entries occur in a fixed order, but not necessarily in exactly the same columns) and that there are no missing data values in the file. The control instruction (METHOD=ML) is the specification of the estimation procedure to be used by EQS in this run (ML is the abbreviation for Maximum Likelihood – see Chapter 3).

If the data are to be read as a covariance or correlation matrix, then the MATRIX statement is replaced by:

MATRIX=COVARIANCE;

(or COV, if abbreviated) or

MATRIX=CORRELATION;

(or COR, if abbreviated). In some analyses you may wish to analyse a correlation matrix, although the data might have been read into EQS, for example, as a covariance matrix. In this situation you use the additional statement:

ANALYSIS=COR;

Alternatively, you may wish to read the data in the form of a correlation matrix together with the standard deviations of the variables (see below). In this case, unless you specify otherwise, EQS will automatically analyse a matrix of variances and covariances. For your own information, however, you may wish to be explicit and include the statement:

ANALYSIS=COV;

/LABELS (Optional)

This is for an optional paragraph in which you give sensible and easily interpretable names for your variables. It can be used for the identification of both observed variables and for hypothetical latent variables (or factors). In EQS the observed variables are automatically given the names V1, V2, . . . , V*n*, solely according to the column order in which they appear in the data. Reading the data from Display 1.1, for example, V1 would be the length measured by a ruler, V2 Graham's guess, V3 Brian's guess and V4 Andrew's guess. In fitting a simple factor analysis model to these data one might specify a single latent variable. This would be called F1 by EQS. The use of the /LABELS paragraph is illustrated in Display 2.4. Here the labelling should be obvious. We have used the label RULER for the so-called true length to distinguish it from the latent variable (factor) which we call LENGTH. Labels may be one to eight characters in length and may only be used to identify observed variables (Vs) or latent variables or common factors (Fs). Labels may be given in any order and all on one line, but remember that they should be separated by the mandatory semicolons.

```
/LABELS
 V1=RULER;
 V2=GRAHAM;
 V3=BRIAN;
 V4=ANDREW;
 F1=LENGTH;
```

DISPLAY 2.4 An example of the use of the /LABELS paragraph.

/EQUATIONS

The /EQUATIONS keyword is followed by detailed information concerning the particular model to be fitted by EQS. The model is specified by one or more equations. One, and only one, equation is required for each dependent variable, and the dependent variables may be either observed (the Vs) or latent (the Fs).

EQS uses four types of variable. We have already referred to measured variables and latent variables (factors) which are referred to by the characters 'V' and 'F', respectively. Residuals of observed variables are denoted by 'E', and the residuals or disturbances of latent variables are denoted by 'D'. The use of the four letters V, F, E and D is summarized in Display 2.5 and will be clarified below by their use in examples.

Returning to the string data, a single factor model for all four measurements is described as in Display 2.6. Each observed variable, V1 to V4, is regressed on, or loads on, the factor F1. The residuals for this regression are called E1 to E4, respectively. The parameters within each of the equations may be specified as either fixed (i.e. specified constants) or free (to be estimated from the data). The '1*' before F1 in each of the last three equations

Code	*Name*	*Meaning*
V	Variable	Measured variable
F	Factor	Latent variable
E	Error	Residual of V
D	Disturbance	Residual of F

DISPLAY 2.5 Types of variable in EQS.

```
/EQUATIONS
 V1=1F1 + E1;
 V2=1*F1 + E2;
 V3=1*F1 + E3;
 V4=1*F1 + E4;
```

DISPLAY 2.6 An example of the /EQUATIONS paragraph.

is telling EQS that the corresponding regression coefficient or loading is a free parameter to be estimated (conveyed by the '*') and that its starting value is 1. In the case of the ruler measurement (V1) the regression coefficient has been constrained to be *fixed* at 1 (i.e. there is no *). A starting value is needed for each of the parameters to be estimated because the fitting algorithm involves iterative estimation, starting from a suitable approximation to the required results and proceeding to their 'optimum' values. Any suitable number could have been specified for the starting values for the regression coefficients, although for numerical reasons it is often wise to choose one which is fairly close to the suspected solution. The starting values can, for example, be chosen from prior knowledge, preliminary graphical analysis, or from a previous EQS run of a (perhaps) simpler model.

/VARIANCES

This keyword is used to introduce a paragraph in which we provide specifications for the variances of *independent* variables. As in the case of regression coefficients, these variances can be fixed in value or can be free to be estimated from the data. Independent variables are defined in EQS as those measured or latent variables (Vs or Fs) that are never regressed on any other variables in the system of specified equations (that is, they *never* appear to the left of the equality sign in any of the equations in the /EQUATIONS paragraph). If they are *ever* regressed on any other variable than they are referred to as being *dependent* by EQS, even though they may appear to the right of the equality sign in one or more of the other specified equations. Only V and F variables can be dependent variables but not all of them need necessarily be so. E and D variables must always appear to the right of the equality sign; that is, they are always independent variables.

In the factor analysis model for the four string measures there are four dependent variables (V1, V2, V3 and V4). There are five independent variables (F1, E1, E2, E3 and E4). Accordingly, in the /VARIABLES paragraph we need to specify four variances. For example:

```
/VARIANCES
F1=3*;
E1 TO E4=0.1*;
```

Here, the starting values are guesses based on preliminary examination of the data. The last line of this /VARIANCES paragraph could have been specified as four separate statements:

```
E1=0.1*; E2=0.1*; E3=0.1*; E4=0.1*;
```

This would have allowed specification of different starting values. Alternatively, we could have used:

```
E1,E2,E3,E4=0.1*;
```

/COVARIANCES (Optional)

In our string example we have chosen statements in the /EQUATIONS and /VARIANCES paragraphs which jointly specify which parameters are fixed (constrained) and which are free to be estimated. A /COVARIANCES (optional) paragraph could also be used to specify covariation among independent variables (but *not* dependent variables). For example:

```
/COVARIANCES
E2,E3=0.2*;
```

This specifies that the covariance between Graham's and Brian's errors is a free parameter with starting value 0.2. Brian may have seen Graham's guesses prior to making his own – so introducing the possibility of correlated measurement errors. Other correlated independent variables will be introduced in Chapter 3.

Returning to our /EQUATIONS and /VARIANCES paragraphs, we could have chosen to parametrize the model in an alternative way (see Sections 1.4 and 1.5). We could have fixed the variance of F1 at 1.0 (or any other arbitrary constant). In doing this we would have had to release the constraint on the regression coefficient for the ruler measurement – allowing it to be freely estimated. Which of the two versions of constraint is chosen is completely arbitrary and simply a matter of tradition or personal taste.

/CONSTRAINTS (Optional)

Here we can introduce either simple equality constraints or general linear constraints. We will limit our explanation to the statements of simple equality. In our factor analysis example we might wish to test whether the precisions (measurement error variances) of our three raters are equal. Similarly, we may wish to constrain the corresponding regression coefficients to be equal. These constraints would be achieved in the following way:

```
/CONSTRAINTS
(E2,E2)=(E3,E3)=(E4,E4);
(V2,F1)=(V3,F1)=(V4,F1);
```

Here (E2,E2), (E3,E3) and (E4,E4) are the variances of E2, E3 and E4, respectively. (V2,F1), (V3,F1) and (V4,F1) are the regression coefficients for the regression of V2 on F1, V3 on F1 and V4 on F1, respectively. **Note the mandatory use of brackets!** Note, also, that starting values for parameter estimates must be consistent with any constraints imposed. If you introduce an equality constraint for two or more parameters, for example, then each of these parameters must have the same starting value.

/MATRIX (Optional)

This paragraph is needed when data are provided in the form of either a covariance matrix or a correlation matrix. It is not used when raw data are input. Do not confuse the /MATRIX keyword with the MATRIX= statement under the /SPECIFICATIONS keyword, although the two statements, if both are used, must be used consistently. You will be aware, we hope, that both covariance and correlation matrices are symetrical about their main diagonal. This implies that if the entries to the lower left of this diagonal are known then there is no need to provide those to the top right. Hence the full symetrical matrix is frequently summarized by a lower-triangular matrix. This is best clarified by an example. Taking the string data, it could be input in the form of variances and covariances as in Display 2.7(a). Note that there are spaces between the entries on any particular line (the data all being read as free format) and there are *no semicolons at the ends of the lines*. 3.878, 2.121, 2.655 and 3.235 are the variances of the ruler measurement, Graham's guess, that of Brian and that of Andrew, respectively. 2.811 is the covariance of the ruler and Graham's guess, and so on. The equivalent lower triagular correlation matrix is given in Display 2.7(b), where in this case, the diagonal entries are all 1s.

(a) A variance-covariance matrix in lower-triangular form

```
/MATRIX
3.878
2.811  2.121
3.148  2.267  2.655
3.506  2.569  2.838  3.235
```

(b) A correlation matrix in lower-triangular form

```
/MATRIX
1.000
0.9802  1.000
0.9811  0.9553  1.000
0.9899  0.9807  0.9684  1.000
```

DISPLAY 2.7 Examples of the /MATRIX paragraph.

/STANDARD DEVIATIONS (Optional)

If the input matrix is a correlation matrix rather than a covariance matrix, the correlations are usually transformed to covariances before the analysis. In order for EQS to do this, it is clearly necessary to provide it with the standard deviations of the variables ($\text{Cov}(A,B) = \text{Corr}(A,B).\text{sd}(A).\text{sd}(B)$ – see Exercise 1.1). For the string measurements this would require the following:

```
/STANDARD DEVIATIONS
1.9692   1.4563   1.6294   1.7987
```

where the standard deviations are read in the order V1 to V4 as expected. Note, again, that there is no semicolon at the end of the line.

Means are not relevant to most covariance structure models, but if needed they can be input through the use of a paragraph identified by the keyword /MEANS. This will be illustrated in Chapter 3.

2.5 PUTTING IT ALL TOGETHER

An example of an input file containing the instructions for a simple factor analysis of the string data is shown in Display 2.8.

```
/TITLE
 Single Factor Model
 Data are guesses of string lengths
 From Display 1.1
/SPECIFICATION
 CASES=15;
 VARIABLES=4;
 METHOD=ML;
 MATRIX=CORRELATION;
 ANALYSIS=COVARIANCE;
/LABELS
 V1=RULER;
 V2=GRAHAM;
 V3=BRIAN;
 V4=ANDREW;
 F1=LENGTH;
/EQUATIONS
 V1=1F1 + E1;
 V2=1*F1 + E2;
 V3=1*F1 + E3;
 V4=1*F1 + E4;
/VARIANCES
 F1=3.0*;
 E1 TO E4=0.1*;
/MATRIX
 1.000
 0.9802  1.000
 0.9811  0.9553  1.000
 0.9899  0.9807  0.9684  1.000
/STANDARD DEVIATIONS
 1.9692  1.4563  1.6294  1.7987
/END
```

DISPLAY 2.8 EQS input file for a single factor model for string data in Display 1.1.

The data are read in as a combination of a correlation matrix and a row of standard deviations. From these EQS can calculate the covariances which are to be analysed as indicated by the ANALYSIS=COVARIANCE command. Note that the input file is terminated by one further keyword:

/END

Exercise 2.1

Consider the following /EQUATIONS paragraph:

```
/EQUATIONS
V1=F1+E1;
V2=1*F1+E2;
V3=1*F2+E3;
V4=1*F2+E4;
F2=1*F1+D2;
```

How many independent variables are there? How many dependent variables are there? Which are the independent variables? Draw a path diagram to illustrate the model describe by these five equations.

Exercise 2.2

Interpret the contents of Display 2.1. Create and run this EQS job and interpret the output.

2.6 THE OUTPUT FILE

Running the program input file shown in Display 2.8 produces eight pages of EQS output. The contents of the first two pages are given in Display 2.9. Page 1 lists the lines of the input file and numbers them; then there is a simple statement saying that 32 records (lines) of an input file have been read. Page 2 begins with the first line of the /TITLE paragraph, together with the date of the run. It then proceeds to describe the analysis that has been requested in the input program. Here we see the required covariance matrix. The summary entitled 'BENTLER-WEEKS STRUCTURAL REPRESENTATION' simply describes the number and identity of both the dependent and independent variables in the model (using the definitions of dependent and independent as described in the explanation of the /EQUATIONS paragraph in Section 2.4, above). At the bottom of this page there is the statement: 'DETERMINANT OF INPUT MATRIX IS .17450D-02'. All is well as long as this determinant is greater than zero.

Display 2.10 gives pages 3 and 4 of the EQS output. These, having started with the statement that maximum likelihood estimates are being used under the assumption of multivariate nor-

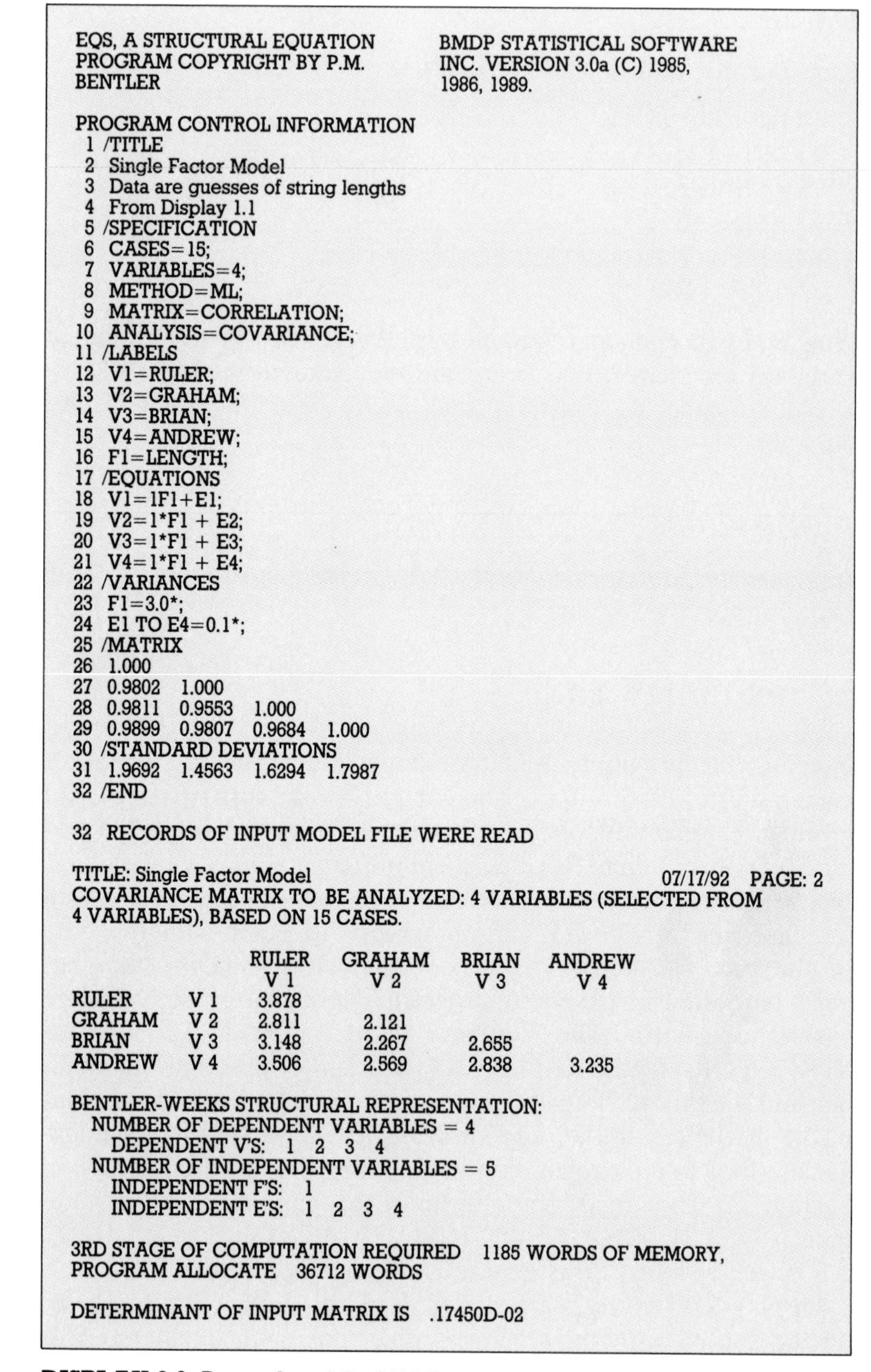

```
EQS, A STRUCTURAL EQUATION          BMDP STATISTICAL SOFTWARE
PROGRAM COPYRIGHT BY P.M.           INC. VERSION 3.0a (C) 1985,
BENTLER                             1986, 1989.

PROGRAM CONTROL INFORMATION
 1 /TITLE
 2  Single Factor Model
 3  Data are guesses of string lengths
 4  From Display 1.1
 5 /SPECIFICATION
 6  CASES=15;
 7  VARIABLES=4;
 8  METHOD=ML;
 9  MATRIX=CORRELATION;
10  ANALYSIS=COVARIANCE;
11 /LABELS
12  V1=RULER;
13  V2=GRAHAM;
14  V3=BRIAN;
15  V4=ANDREW;
16  F1=LENGTH;
17 /EQUATIONS
18  V1=1F1+E1;
19  V2=1*F1 + E2;
20  V3=1*F1 + E3;
21  V4=1*F1 + E4;
22 /VARIANCES
23  F1=3.0*;
24  E1 TO E4=0.1*;
25 /MATRIX
26  1.000
27  0.9802  1.000
28  0.9811  0.9553  1.000
29  0.9899  0.9807  0.9684  1.000
30 /STANDARD DEVIATIONS
31  1.9692  1.4563  1.6294  1.7987
32 /END

32  RECORDS OF INPUT MODEL FILE WERE READ

TITLE: Single Factor Model                               07/17/92   PAGE: 2
COVARIANCE MATRIX TO BE ANALYZED: 4 VARIABLES (SELECTED FROM
4 VARIABLES), BASED ON 15 CASES.

                     RULER     GRAHAM     BRIAN     ANDREW
                      V 1        V 2       V 3        V 4
RULER        V 1     3.878
GRAHAM       V 2     2.811      2.121
BRIAN        V 3     3.148      2.267     2.655
ANDREW       V 4     3.506      2.569     2.838      3.235

BENTLER-WEEKS STRUCTURAL REPRESENTATION:
   NUMBER OF DEPENDENT VARIABLES = 4
      DEPENDENT V'S:  1  2  3  4
   NUMBER OF INDEPENDENT VARIABLES = 5
      INDEPENDENT F'S:  1
      INDEPENDENT E'S:  1  2  3  4

3RD STAGE OF COMPUTATION REQUIRED   1185 WORDS OF MEMORY,
PROGRAM ALLOCATE   36712 WORDS

DETERMINANT OF INPUT MATRIX IS  .17450D-02
```

DISPLAY 2.9 Pages 1 and 2 of EQS output generated by EQS input file in Display 2.8.

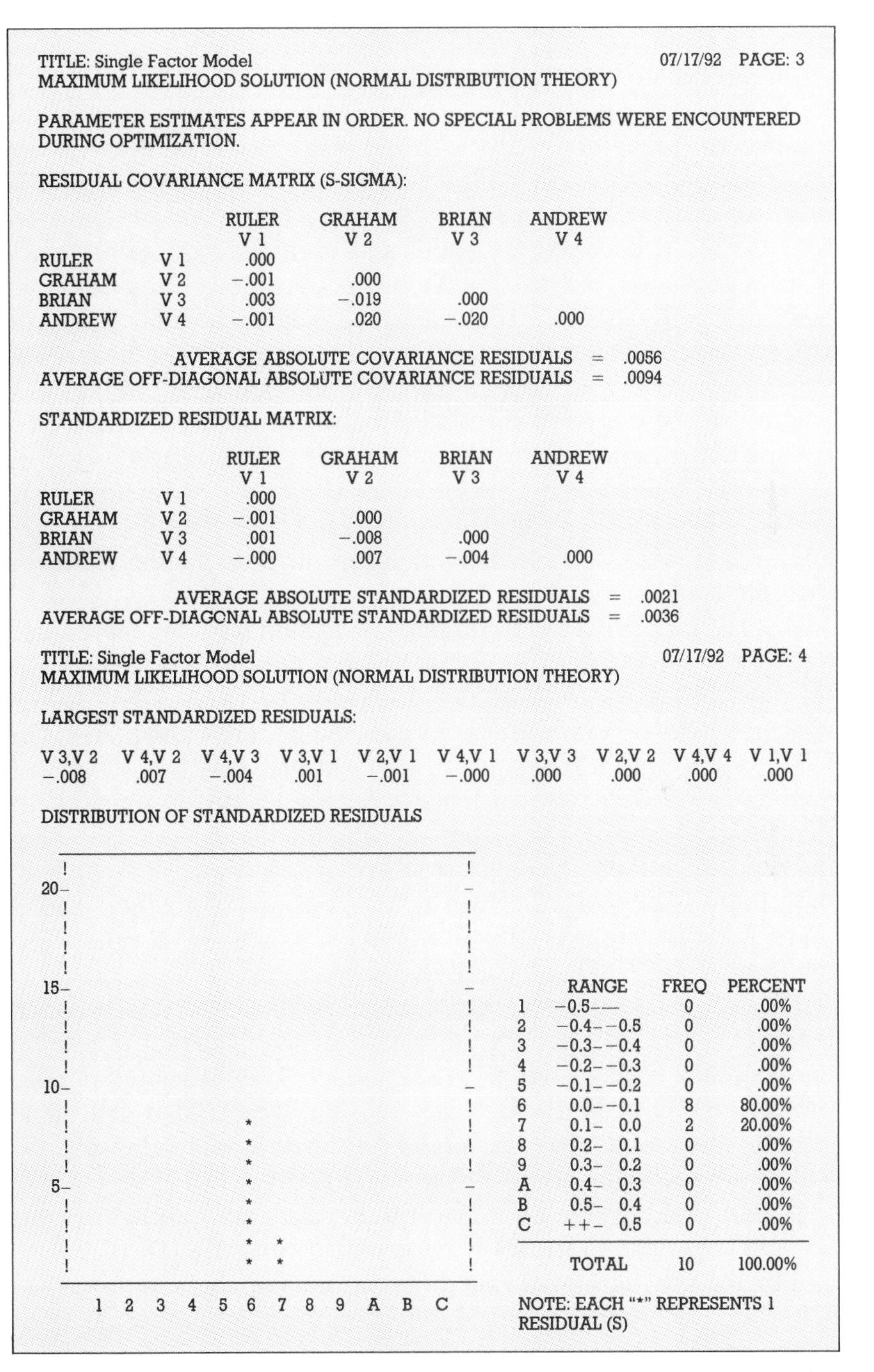

```
TITLE: Single Factor Model                                          07/17/92   PAGE: 3
MAXIMUM LIKELIHOOD SOLUTION (NORMAL DISTRIBUTION THEORY)

PARAMETER ESTIMATES APPEAR IN ORDER. NO SPECIAL PROBLEMS WERE ENCOUNTERED
DURING OPTIMIZATION.

RESIDUAL COVARIANCE MATRIX (S-SIGMA):

                      RULER      GRAHAM     BRIAN      ANDREW
                      V 1        V 2        V 3        V 4
RULER       V 1       .000
GRAHAM      V 2      -.001       .000
BRIAN       V 3       .003      -.019       .000
ANDREW      V 4      -.001       .020      -.020       .000

                AVERAGE ABSOLUTE COVARIANCE RESIDUALS    =   .0056
AVERAGE OFF-DIAGONAL ABSOLUTE COVARIANCE RESIDUALS    =   .0094

STANDARDIZED RESIDUAL MATRIX:

                      RULER      GRAHAM     BRIAN      ANDREW
                      V 1        V 2        V 3        V 4
RULER       V 1       .000
GRAHAM      V 2      -.001       .000
BRIAN       V 3       .001      -.008       .000
ANDREW      V 4      -.000       .007      -.004       .000

                AVERAGE ABSOLUTE STANDARDIZED RESIDUALS    =   .0021
AVERAGE OFF-DIAGONAL ABSOLUTE STANDARDIZED RESIDUALS    =   .0036

TITLE: Single Factor Model                                          07/17/92   PAGE: 4
MAXIMUM LIKELIHOOD SOLUTION (NORMAL DISTRIBUTION THEORY)

LARGEST STANDARDIZED RESIDUALS:

V 3,V 2  V 4,V 2  V 4,V 3  V 3,V 1  V 2,V 1  V 4,V 1  V 3,V 3  V 2,V 2  V 4,V 4  V 1,V 1
 -.008     .007    -.004     .001    -.001    -.000     .000     .000     .000     .000

DISTRIBUTION OF STANDARDIZED RESIDUALS

   ----------------------------------------
   !                                      !
20-                                       -
   !                                      !
   !                                      !
   !                                      !
   !                                      !
15-                                       -        RANGE        FREQ   PERCENT
   !                                      !     1  -0.5--         0       .00%
   !                                      !     2  -0.4--  -0.5   0       .00%
   !                                      !     3  -0.3--  -0.4   0       .00%
   !                                      !     4  -0.2--  -0.3   0       .00%
10-                                       -     5  -0.1--  -0.2   0       .00%
   !                                      !     6   0.0--  -0.1   8     80.00%
   !                 *                    !     7   0.1-    0.0   2     20.00%
   !                 *                    !     8   0.2-    0.1   0       .00%
   !                 *                    !     9   0.3-    0.2   0       .00%
 5-                  *                    -     A   0.4-    0.3   0       .00%
   !                 *                    !     B   0.5-    0.4   0       .00%
   !                 *                    !     C   ++-     0.5   0       .00%
   !                 *  *                 !    ----------------------------------
   !                 *  *                 !          TOTAL         10    100.00%
   ----------------------------------------
      1  2  3  4  5  6  7  8  9  A  B  C        NOTE: EACH "*" REPRESENTS 1
                                                RESIDUAL (S)
```

DISPLAY 2.10 Pages 3 and 4 of EQS output generated by EQS input file in Display 2.8.

mality of the Vs and Es, produce various summaries of the residual covariance matrix. If we represent the observed covariance matrix by $\boldsymbol{S}$ (with elements s_{ij}) and the covariance matrix predicted by the parameter estimates by $\hat{\boldsymbol{\Sigma}}$ (with elements $\hat{\sigma}_{ij}$) then the residual covariance matrix is the difference, $\boldsymbol{S}-\hat{\boldsymbol{\Sigma}}$. The elements of this residual matrix are then simply $s_{ij}-\hat{\sigma}_{ij}$. The values of these residuals should be relatively small and evenly spread among variables if the model is a reasonable one for the data. Large residuals associated with specific variables are an indication of poor fit. The output also includes summaries of standardized residuals. The standardized residual matrix contains the elements $r_{ij}-(\hat{\sigma}_{ij}/s_{ii}s_{jj})$, where r_{ij} is the observed correlation between variables i and j, σ_{ij} is the predicted covariance, and s_{ii} and s_{jj} are their observed standard deviations. The standardized residuals (based on correlations) are easier to interpret than the unstandardized ones (based on covariances) since they are not dependent on the scale of the observed measurements.

Display 2.11 provides a goodness-of-fit summary for the model, together with details of the iterations required to fit the model to the data. Here we restrict the discussion to two goodness-of-fit statistics; the others will be introduced in later chapters. The independence model chi-squared (χ^2) statistic (here 148.520 with 6 degrees of freedom) provides a significance test for the hypothesis that the four observed variables are all mutually independent. If χ^2 indicates that the observed variables are, indeed, independent, then there is no point in investigating the output further! Either there is no covariance structure to be explored or, alternatively, the sample size is far too small to reveal any.

The next statement to consider is 'CHI-SQUARE = 2.347 BASED ON 2 DEGREES OF FREEDOM'. This is the test for the goodness-of-fit of the proposed model. The associated p-value (here 0.30921) is well above any sensible threshold which might indicate lack of fit. Thus the model fits the observed data well. We will return to chi-squared statistics in Chapter 3.

Display 2.12 gives the final three pages of output. Having decided that our model has been specified correctly (Display 2.9), and that it provides a reasonable description of the data (Displays 2.10 and 2.11) we now proceed to look at the more interesting parts of the output – the parameter estimates and their standard errors. Display 2.12 provides details for the regression coefficients. Each of the four equations gives the user's label for the corresponding

```
TITLE: Single Factor Model                                   07/17/92   PAGE: 5
MAXIMUM LIKELIHOOD SOLUTION (NORMAL DISTRIBUTION THEORY)

GOODNESS OF FIT SUMMARY

INDEPENDENCE MODEL CHI-SQUARE = 148.520, BASED ON 6 DEGREES OF
FREEDOM

INDEPENDENCE AIC = 136.52010     INDEPENDENCE CAIC = 126.27180
       MODEL AIC = -1.65251             MODEL CAIC = -5.06861

CHI-SQUARE = 2.347 BASED ON 2 DEGREES OF FREEDOM
PROBABILITY VALUE FOR THE CHI-SQUARE STATISTIC IS .30921
THE NORMAL THEORY RLS CHI-SQUARE FOR THIS ML SOLUTION IS 2.535.

BENTLER-BONETT NORMED        FIT INDEX = .984
BENTLER-BONETT NONNORMED FIT INDEX = .993
COMPARATIVE                  FIT INDEX = .998

            ITERATIVE SUMMARY

            PARAMETER
ITERATION   ABS CHANGE    ALPHA     FUNCTION
1             .192741    1.00000    1.03247
2             .053643    1.00000     .19886
3             .006345    1.00000     .17069
4             .002189    1.00000     .16803
5             .000547    1.00000     .16768
```

DISPLAY 2.11 Page 5 of EQS output generated by EQS input file in Display 2.8.

observed variables (V1 to V4). To the right of the second equality sign are given the regression coefficient estimates. The top one (without an asterisk) has a value of 1.000 and, in fact, is not an estimate. It was constrained to be 1 by our model. That for Graham's guess is 0.727. (Graham is consistently underestimating the length of the pieces of string; the degree of underestimation, on average, increasing with increasing string length). The precision of this estimate is provided by an estimate of its standard error. This is given immediately below the estimate, and here it is 0.039. The third number in this column (18.710) is the ratio of the regression coefficient estimate to its standard error (that is, 0.727/0.039). On the assumption that we have fitted an appropriate model, this ratio can be regarded as a test statistic for the null

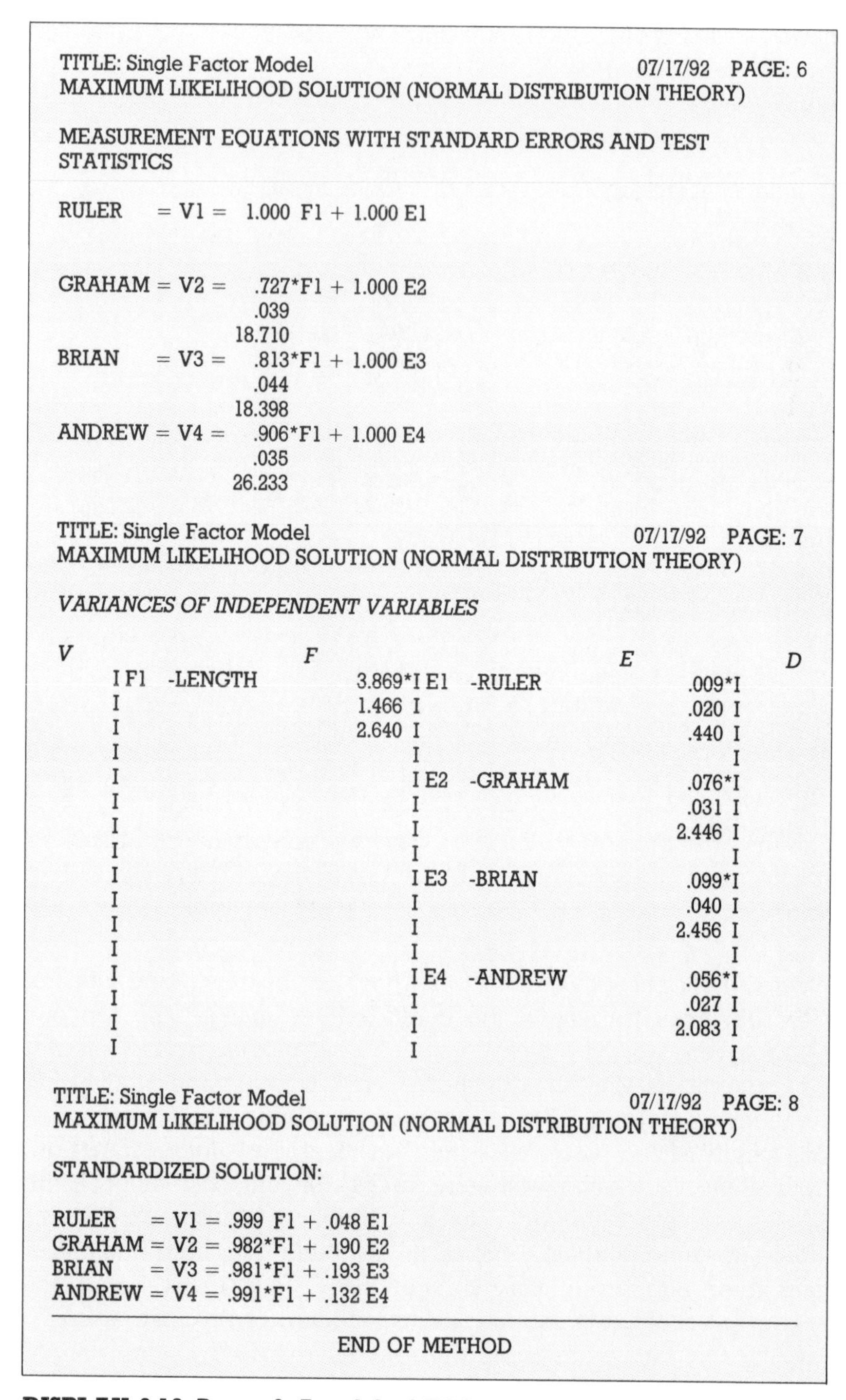

```
TITLE: Single Factor Model                                   07/17/92   PAGE: 6
MAXIMUM LIKELIHOOD SOLUTION (NORMAL DISTRIBUTION THEORY)

MEASUREMENT EQUATIONS WITH STANDARD ERRORS AND TEST
STATISTICS

RULER    = V1 =  1.000 F1 + 1.000 E1


GRAHAM = V2 =    .727*F1 + 1.000 E2
                 .039
               18.710
BRIAN    = V3 =  .813*F1 + 1.000 E3
                 .044
               18.398
ANDREW = V4 =    .906*F1 + 1.000 E4
                 .035
               26.233

TITLE: Single Factor Model                                   07/17/92   PAGE: 7
MAXIMUM LIKELIHOOD SOLUTION (NORMAL DISTRIBUTION THEORY)

VARIANCES OF INDEPENDENT VARIABLES

V                        F                             E                    D
      I F1  -LENGTH          3.869*I E1   -RULER              .009*I
      I                      1.466 I                          .020 I
      I                      2.640 I                          .440 I
      I                            I                               I
      I                            I E2   -GRAHAM             .076*I
      I                            I                          .031 I
      I                            I                         2.446 I
      I                            I                               I
      I                            I E3   -BRIAN              .099*I
      I                            I                          .040 I
      I                            I                         2.456 I
      I                            I                               I
      I                            I E4   -ANDREW             .056*I
      I                            I                          .027 I
      I                            I                         2.083 I
      I                            I                               I

TITLE: Single Factor Model                                   07/17/92   PAGE: 8
MAXIMUM LIKELIHOOD SOLUTION (NORMAL DISTRIBUTION THEORY)

STANDARDIZED SOLUTION:

RULER    = V1 = .999 F1 + .048 E1
GRAHAM = V2 = .982*F1 + .190 E2
BRIAN    = V3 = .981*F1 + .193 E3
ANDREW = V4 = .991*F1 + .132 E4
--------------------------------------------------------------------------
                           END OF METHOD
```

DISPLAY 2.12 Pages 6, 7 and 8 of EQS output generated by EQS input file in Display 2.8.

hypothesis that the given parameter is zero. These ratios are the familiar large-sample *z*-tests for normally distributed random variables. These tests, however, should be used with care and other, better, approaches will be introduced later in the primer. In the present example, however, the test is of little interest in any case. Finally, to the right of the measurement equation for Graham's guesses is '+ 1.000 E2'. Here, again there is no parameter estimate to be given, the '1.00' (without an asterisk) simply indicating a constant which EQS has fixed to be 1.

Page 7 (in display 2.12) provides the estimates for the variances of the independent variables. These are presented in four columns: V, F, E, and D. There are no independent Vs in our model, so this column is empty. There are also no Ds in the model, therefore this column is also empty. The estimate of the variance of F1 is given as 3.869 with an estimated standard error of 1.466. The corresponding *z*-statistic is 2.640 (3.869/1.466). Moving on to the variances of the error terms or residuals, these are estimated by 0.009, 0.076, 0.099 and 0.056 for the ruler, Graham's guesses, Brian's guesses and Andrew's guesses, respectively. Not surprisingly, that for the ruler is considerably lower than those for the guesses! Andrew's guesses appear to be more precise than the other two, and perhaps Brian's are the worst.

The final section of Display 2.12 (page 8 at the output file) provides the so-called 'STANDARDIZED SOLUTION'. The parameter estimates have here been rescaled following standardization of both the latent variable (F1) and the residuals (E1 to E4) to have unit variance. Consider the ruler, for example:

$$\text{Ruler} = \text{V1} = 0.999\ \text{F1} + 0.048\ \text{E1}$$

and therefore, if Var(F1)=Var(E1)=1, then:

$$\begin{aligned}\text{Var(V1)} &= 0.999^2 + 0.048^2\\ &= 1\end{aligned}$$

The communality (an example of factor analysis terminology indicating the proportion of the variance of V1 explained by the variance of the common factors, here F1 – see Section 1.5) or, equivalently (in this example), the reliability of the ruler is 0.999^2. This is, of course, the same as $1-0.048^2$. Similarly:

$$\begin{aligned}\text{Var(V2)} &= 0.982^2 + 0.190,\\ \text{Var(V3)} &= 0.981^2 + 0.193\\ \text{Var(V4)} &= 0.991^2 + 0.132\end{aligned}$$

The corresponding reliabilities are 0.982^2, 0.981^2 and 0.991^2, respectively.

Exercise 2.3

Using the EQS input file given in Display 2.8, modify the first line of the /EQUATIONS paragraph to:

V1=1*F1+E1;

In addition, modify the first line of the /VARIANCES paragraph to:

F1=1.0;

Rerun the EQS job and interpret the output with reference to that provided in Displays 2.9–2.12.

Exercise 2.4

Modify the input file give in Display 2.6 in order to produce a simple factor model for only two of the measured variables: V1 (Graham's guess) and V2 (Brian's guess). Using a single common factor (F1) with a variance fixed at 1.0, estimate the regression coefficients for V1 and V2 under the constraint that they are equal, together with the constraint that the variances of residuals are also equal (see Section 1.4).

Exercise 2.5

Returning to Display 2.8, drop the first line for the /EQUATIONS paragraph. In addition, change the second line of the /VARIANCES paragraph to:

E2 to E4=0.1*;

We now have the just identified factor model for three variables which was described in Section 1.5. Run this EQS job and interpret the output. Pay particular attention to the goodness-of-fit summary and to the residuals.

2.7 SUMMARY

We have described the basic commands (keywords and associated paragraphs) in EQS and have illustrated how they are put together to produce on EQS input or command file. Having run a particular EQS job (fitting a sample factor analysis model to the string data). We then explained how to read and interpret the major components of the output file. You should now be in a position to start running EQS jobs, at least for the simpler factor analysis models, yourself. The next chapter will introduce you to various model-fitting strategies.

Statistical modelling in EQS

3

3.1 INTRODUCTION

In this chapter we start by reviewing the fitting methods available in EQS for over-identified models. We also discuss global goodness-of-fit statistics and their associated chi-squared tests. Particular attention is paid to testing the equality of two or more parameter values through the introduction of constraints. Ideas will be illustrated using the already familiar string data. We will also analyse data for a simple test-retest reliability study of a psychiatric questionnaire. This example will be used to illustrate a **confirmatory factor analysis** model involving two correlated factors and also an equivalent **structural equation model**.

The chapter is completed by discussion of two commonly occurring problems. The first arises through trying to fit under-identified models. The second comes from indications of impermissible parameter values (correlations greater than unity, for example, or negative variance estimates). The latter are usually automatically constrained by the EQS program to take values on the boundary between the permissible and impermissible values.

After reading this chapter you will be in a position to proceed to try fitting more ambitious models to more complex data sets.

3.2 ESTIMATION AND GOODNESS OF FIT

Using the terminology first introduced in Section 2.6, the matrix S is the observed covariance matrix (i.e. the data) and Σ is the covariance matrix predicted using the model parameters. If the parameters of any given model are represented by the vector $\boldsymbol{\theta}'=(\theta_1, \theta_2, \ldots, \theta_t)$, where t is the number of parameters to be estimated, then we can be more explicit in our description of a particular predicted covariance matrix by using the term $\boldsymbol{\Sigma}(\boldsymbol{\theta})$. The elements of the matrices $\boldsymbol{S}$ and $\boldsymbol{\Sigma}$ are represented by $\{s_{ij}\}$ and $\{\sigma_{ij}\}$ respectively. Estimates for the latter are represented by $\{\hat{\sigma}_{ij}\}$.

We also saw in Section 2.6 that the residual terms $s_{ij} - \hat{\sigma}_{ij}$ are used to evaluate the fitted model. Our general aim is to find estimates of $\boldsymbol{\theta}$ (and hence $\boldsymbol{\Sigma}(\boldsymbol{\theta})$) that minimize some function of these residuals. That is, we wish to minimize some function of $\boldsymbol{S}-\boldsymbol{\Sigma}(\boldsymbol{\theta})$. There are three commonly used estimation methods in situations in which we are prepared to assume multivariate normality of the observed measurements. The first is based on minimizing the sum of the squared residuals – **ordinary least squares** or (OLS). The second is based on minimizing a weighted combination of the squared residuals with fixed weights based on the observed covariance matrix $\boldsymbol{S}$. This is known as **generalized least squares** (GLS). The third option is a modification of GLS in which the weights are updated at each iteration using the current estimates of the covariance matrix $\boldsymbol{\Sigma}(\boldsymbol{\theta})$. This **reweighted least squares** (RLS) method can be shown to be equivalent to the method of **maximum likelihood** (ML) in the case of multivariate normality. EQS also provides estimation procedures based on distributional assumptions other than normality, but these will not be discussed further here; they will be introduced in Chapter 8.

Now, if the discrepancy function to be minimized in OLS, GLS or ML(RLS) is represented by $F(\boldsymbol{S},\boldsymbol{\Sigma}(\boldsymbol{\theta}))$, it is sensible to use the optimal value, $F(\boldsymbol{S},\boldsymbol{\Sigma}(\boldsymbol{\theta}))$, as a measure of the goodness of fit of the model: the lower the value the better the fit. In fact, assuming multivariate normality, it is possible to show that, for GLS and ML, $F(\boldsymbol{S},\boldsymbol{\Sigma}(\boldsymbol{\theta}))$ has (asymptotically) a chi-squared distribution under the null hypothesis that the covariance matrix is of the form predicted by the model. The degrees of freedom for the chi-squared distribution is given by:

$$\nu = p(p + 1)/2 - t$$

where p is the number of measured variables and t the number of estimated (free) parameter values. The term 'asymptotically' can be interpreted as meaning 'for sufficiently large N' – N being the sample size.

Model (a)

V1=F1 + E1;
No constraints
(see Display 2.8)

Model (b)

V1=F1;
(i.e. Ruler is the 'truth')
! Remember to delete E1 from /VARIANCES paragraph

Model (c)

V1=F1;
/CONSTRAINTS
(E2,E2)=(E3,E3)=(E4,E4);

Model (d)

V1=F1;
/CONSTRAINTS
(E2,E2)=(E3,E3)=(E4,E4);
(V2,F1)=(V3,F1)=(V4,F1);

Model (e)

V1=F1 + E1;
/CONSTRAINTS
(E2,E2)=(E3,E3)=(E4,E4);

DISPLAY 3.1 Fitting a series of models to the string data.

3.3 TESTING CONSTRAINTS

Returning to the string data and to the EQS program shown in Display 2.8 (model (a) in Display 3.1), we now test a few simple and obvious hypotheses. The first one is based on the assumption that the measurements made through the use of the ruler are, indeed, error-free. The appropriate constraint here is to fix the variance of E1 at zero. The simplest way of doing this is, in fact, to drop any mention of E1 for the model defined by the first line of the /EQUATIONS paragraph (that is, to specify that V1=F1). This is model (b) of Display 3.1 and the effect of changing the

Model	*Chi-squared*	*d.f.*
(a)	2.347	2
(b)	2.487	3
(c)	3.113	5
(d)	12.547	7
(e)	3.110	4

DISPLAY 3.2 Goodness of fit for string models.

model is shown by the chi-squared statistic given for model (b) in Display 3.2. Model (b) has one free parameter less (the variance of E1) and therefore has one more degree of freedom (3 instead of the original 2). The chi-squared statistic has increased, but the increase is negligible. A formal test of the null hypothesis that Var(E1)=0 is provided by the differences in the two chi-squared statistics (that is, 2.487 − 2.347), which itself should be distributed as a chi-square variate with degrees of freedom equal to the difference in the degrees of freedom of the two models, that is, 3−2 = 1 in this case. The associated probability is well above 0.05, so the data provide us with insufficient evidence to reject the null hypothesis (although it should be obvious to all readers that the ruler cannot actually provide completely error-free measurements). We will keep the constraint implied by the V1=F1 equation and proceed to compare the performance of the guesses of Graham, Brian and Andrew.

Model (c) in Display 3.1 introduces the constraints to test the equality of the remaining three error variances (Var(E2), Var(E3) and Var(E4)). Here three free parameters have been replaced by a single common parameter – the degrees of freedom, therefore, increase by 2. Looking at Display 3.2, it can be seen that the change in the chi-squared statistic is again neglible. Model (c) provides an excellent fit to the data. We now introduce further constraints to equate the slopes of the calibration lines relating the guesses of the three raters with F1 (that is, their factor loadings for F1). This is illustrated by model (d) in Display 3.1. The degrees of freedom are increased by 2 to yield a model χ^2 statistic with 7 degrees of freedom. Looking at Display 3.2, the goodness-of-fit chi-square statistic has jumped to 12.547. The probability associated with this model is, however, still greater than 0.05 so one might be temped to accept it as a reasonable fit. The change in χ^2 in going from model (c) to model (d) is, however, 9.434 with 2 degrees of freedom. This change is highly significant ($p < 0.01$). The relative biases for the three raters do not appear to be equal.

Finally, consider model (e). Here, the ruler-based measurement is allowed to be fallible, but the error variances of the three guesses are constrained to be equal (the factor loadings, however, are not constrained). Again, the fit is very good. The difference between the chi-squared statistics for models (c) and (e) provide an alternative test for the hypothesis that the ruler is error-free, but here the test is conditional on the equality constraints concerning the error variances for the other three measures. In this example it does not matter at all whether the test is conditional on the other constraints, but in some cases it might. Readers must be careful to consider the order in which they introduce constraints (that is, the exact nature of their hypotheses).

Exercise 3.1

Starting with model (c) of Display 3.1 try constraining the factor loadings for *pairs* of raters (there are three possibilities) and comment on your results.

Exercise 3.2

Repeat the model-fitting procedures illustrated by the section above, but here restricting your models to data on Graham's, Brian's and Andrew's guesses alone (i.e. drop all references to V1 in your /EQUATIONS paragraph).

3.4 MODELS WITH TWO FACTORS

Display 3.3 shows some data on the measurement of psychological distress obtained through the use of Goldberg's (1972) General Health Questionnaire (GHQ). A group of clinical psychology students were asked to complete this questionnaire on two occasions, three days apart. On each occasion the sum of the responses to the odd-numbered items was obtained, as was that for the even-

Odd1	*Even1*	*GHQ1*	*Odd2*	*Even2*	*GHQ2*
7	5	12	7	5	12
4	4	8	4	3	7
12	10	22	12	12	24
5	5	10	7	7	14
3	7	10	3	5	8
4	2	6	3	1	4
3	5	8	2	3	5
2	2	4	3	3	6
7	7	14	6	8	14
3	3	6	2	3	5
0	2	2	3	2	5
11	11	22	8	8	16

Key to variable names:
Odd1 and Odd2 are the sums of the odd items at times 1 and 2, respectively.
Even1 and Even2 are the respective sums for the even items and GHQ1 and GHQ2 are the total scores at the two times.

DISPLAY 3.3 GHQ Scores (sums of odd and even items and their total) on two occasions (Dunn, 1992).

numbered items. GHQ1 and GHQ2 are the overall totals (GHQ scores) for each of the two occasions. Here we will consider two alternative models for the four measurements labelled Odd1, Even1, Odd2 and Even2.

First, it is clear that we should consider these observed measurements as fallible indicators of one or more latent traits. Odd1 and Even1 can be postulated to be indicators of distress at time 1 (Distress1) and, similarly, Odd2 and Even2 can be thought to be fallible indicators of distress at the second occasion (Distress2). It is possible that a subject's distress is completely stable (that is, Distress1 = Distress2), but this is unlikely – there are bound to be day-to-day fluctuations in stress and mood. In our statistical models Distress1 and Distress2 are the labels given to two factors F1 and F2, respectively. See Display 3.4 for a representation of these measurement models.

Now, there is no reason to believe that the precision of the sum of the odd items will differ from that of the even ones. Nor is there any reason to suppose that their precisions will differ on the two occasions. In all models, therefore, we use the constraints

	F1	F2
Odd1	X	0
Even1	X	0
Odd2	0	X
Even2	0	X

In this matrix, X denotes a loading free to be estimated and 0 indicates a loading which is constrained to be zero

F1 and F2 can be related via a simple correlation (a confirmatory factor analysis model) or through a structural equation (a structural equation model or 'causal' model).

DISPLAY 3.4 Structure of the Measurement Models for the GHQ Scores.

Var(Odd1) = Var(Odd2) = Var(Even1) = Var(Even2). We also constrain the factor loading for Odd1 to be identical to that for Even1. Similarly, we constrain that for Odd2 to be the same as the loading for Even2. We do not necessarily constrain the respective loadings to be equal for the two occasions, however. This will depend on the type of model being fitted (see below).

We now consider two statistical models for these data which, although they are specified differently, are entirely equivalent (both providing a goodness-of-fit chi-squared statistic of 10.789 with 6 degrees of freedom). In both we have two factors, F1 and F2, indicating psychological distress at time 1 and at time 2, respectively. F1 is measured by Odd1 and Even1, and F2 is measured by Odd2 and Even2. The difference between the two types of model arises from the postulated relationship between F1 and F2. In fitting a confirmatory factor analysis model we can simply allow F1 and F2 to be correlated – the correlation being a free parameter to be estimated. This is illustrated in Display 3.5. Here the data are provided in the form of a correlation matrix together with the standard deviations of the four measures, V1 to V4. The identities of V1 to V4 are indicated in the /LABELS paragraphs. Note that both F1 and F2 have variances fixed at unity, but that the loadings for F1 are *not* constrained to be equal on the two occasions. This allows for the variability between

```
/TITLE
 Two Factor Model
 Data are GHQ scores from 12 students
 Odd vs Even subtotals on two occasions
/SPECIFICATIONS
 CASES=12;
 VARIABLES=4;
 METHOD=ML;
 MATRIX=CORRELATION;
 ANALYSIS=COVARIANCE;
/LABELS
 V1=ODD1;
 V2=EVEN1;
 V3=ODD2;
 V4=EVEN2;
 F1=DISTRESS1;
 F2=DISTRESS2;
/EQUATIONS
 V1=1*F1+E1;
 V2=1*F1+E2;
 V3=1*F2+E3;
 V4=1*F2+E4;
/VARIANCES
 F1 TO F2=1;
 E1 TO E4=1*;
/COVARIANCES
 F1,F2=0.9*;
/CONSTRAINTS
 (E1,E1)=(E2,E2)=(E3,E3)=(E4,E4);
 (V1,F1)=(V2,F1);
 (V3,F2)=(V4,F2);
/MATRIX
 1.000
 0.8645  1.000
 0.9008  0.7516  1.000
 0.8597  0.8682  0.9075  1.000
/STANDARD DEVIATIONS
 3.5792  3.0189  3.0451  3.1909
/END
```

DISPLAY 3.5 EQS input file for a simple two-factor model.

subjects to vary across occasions. An alternative specification could have fixed the factor loadings to be equal across occasions and allowed the variance of F2 to be free.

Finally, note that there is a /COVARIANCES paragraph to

```
/TITLE
 Structural Equation Model
 Data are GHQ scores from 12 students
 Odd vs Even subtotals on two occassions
/SPECIFICATIONS
 CASES=12;
 VARIABLES=4;
 METHOD=ML;
 MATRIX=CORRELATION;
 ANALYSIS=COVARIANCE;
/LABELS
 V1=ODD1;
 V2=EVEN1;
 V3=ODD2;
 V4=EVEN2;
 F1=DISTRESS1;
 F2=DISTRESS2;
/EQUATIONS
 V1=1*F1+E1;
 V2=1*F1+E2;
 V3=1*F2+E3;
 V4=1*F2+E4;
 F2 =1*F1+D2;
/VARIANCES
 F1=1;
 E1 TO E4=1*;
 D2=1*;
/CONSTRAINTS
 (E1,E1)=(E2,E2)=(E3,E3)=(E4,E4);
 (V1,F1)=(V2,F1)=(V3,F2)=(V4,F2);
/MATRIX
 1.000
 0.8645  1.000
 0.9008  0.7516  1.000
 0.8597  0.8682  0.9075  1.000
/STANDARD DEVIATIONS
 3.5792  3.0189  3.0451  3.1909
/END
```

DISPLAY 3.6 EQS input file for a simple structural equation model.

allow F1 and F2 to be correlated. The covariance between F1 and F2 is given a starting value of 0.9 which, because we have specified that both F1 and F2 have unit variance, in this case is actually a correlation. If, on the other hand, we had allowed the variance of

F2 to be free (as in the second specification) this parameter would have been a covariance.

The above model provides quite a satisfactory description of the data, but does not take into account the essential asymmetry on the relation between F1 and F2. As F1 is a state measured *before* F2, and in some way can be thought to have a causal effect on F2, we could introduce a further equation in the /EQUATIONS paragraph to reflect this. This is done in Display 3.6.

Display 3.6 provides an EQS program to fit a simple example of a **structural equation model**. The structural relationship is represented by the equation F2=1*F1+D2. Here F2 is a combination of two components: that arising from ('caused by') F1 and that arising from all other 'causes' and summarized in a random deviation (D2). Note that F2 is now a dependent variable in the sense used by EQS. It does not, therefore, appear in the /VARIANCES paragraph – this applying only to independent variables. There is also no /COVARIANCE paragraph. Finally, all four factor loadings have been constrained to be equal. As the variance of F2 is predicted by the parameter estimates in the structural equation (that is, the variance of D2 and the regression coefficient predicting F2 from F1) it cannot be constrained. If we had not introduced the constraints for the factor loadings across time, the model would have been unidentified (see Exercise 3.5).

Path diagrams for the two above models are provided in Displays 3.7 and 3.8, respectively.

Exercise 3.3

Fit the model in Display 3.5 and interpret the output. Modify the model by simultaneously allowing the variance of F2 to be estimated and constraining all four factor loadings to be equal (that is, (V1,F1) = (V2,F1) = (V3,F2)=(V4,F2)). Compare the resulting output with that produced by the first run.

Exercise 3.4

Fit the model in Display 3.6 and interpret the output. Compare the results with those produced in Exercise 3.3. Modify the model by simultaneously changing the structural equation model to F2=F1+D2 and relaxing the constraint of equality of the factor loadings over time. Comment on the resulting output.

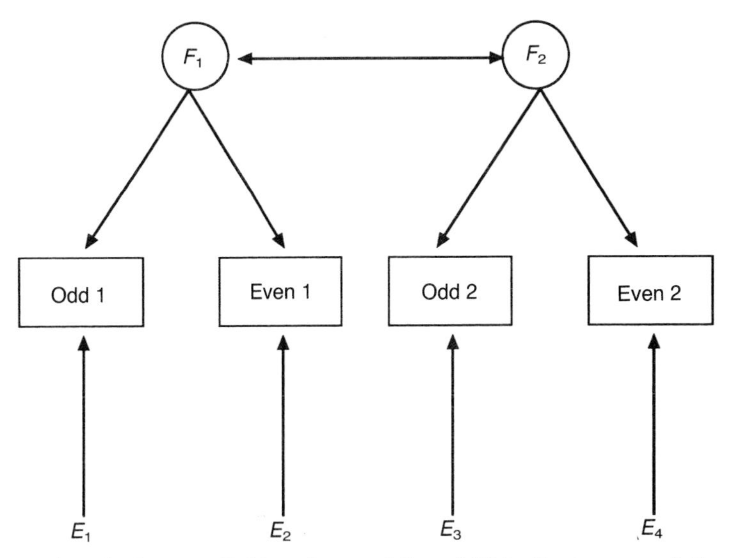

The two-headed arrow linking factors F1 and F2 indicates correlation, without an implied direction of causality.

DISPLAY 3.7 Path diagram for a two-factor model of GHQ scores.

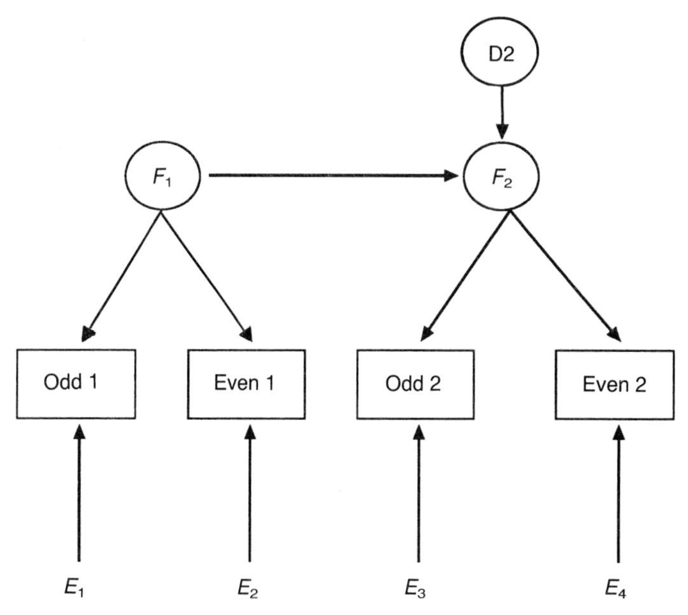

Note that the single-headed arrow linking F2 with F1 implies temporal causality.

DISPLAY 3.8 Path diagram for a structural equation model for GHQ scores.

3.5 TWO COMMONLY OCCURRING PROBLEMS

Before moving on to more challenging data sets we will have one final look at the string data. Readers may have noticed that there is one obvious model missing from Displays 3.1 and 3.2. This is a model in which we acknowledge that the ruler is fallible (that is, we use a model in which we do not constrain the variance of its errors to be zero) and introduce both sets of constraints on the three guesses (that is, both equality of the error variances and equality of the factor loadings). When we do, in fact, fit this model we see a condition code on page 3 of the output file:

```
PARAMETER      CONDITION CODE
E1,E1          CONSTRAINED AT LOWER BOUND
```

This is followed, on page 6, by a warning:

```
TEST RESULTS MAY NOT BE APPROPRIATE DUE TO
CONDITION CODE
```

This implies that, if it were not for constraints provided within EQS, the parameter estimate would not be within the boundaries permitted. In this example the estimate for the variance of E1 would be negative. In the factor analysis literature, the occurrence of a negative estimate for a specific variance is known as a **Heywood case**. Heywood cases can arise through sampling fluctuations (particularly in the case of small samples) or a misspecified model. In either case the result should be treated with care. EQS automatically puts the estimate on the boundary between its permissible and impermissible values (i.e. zero in this case). In the string example, it makes sense to constrain Var(E1) to be zero, but this will not always be so.

Finally, we will fit the model given in Display 2.8, but having first changed the first line of the EQUATIONS paragraph to V1=1*F1+E1. Here we are fitting an unidentified model. EQS quite happily fits this model but, again, provides the warning message. In this case the condition code indicates linear dependence amongst the parameter estimates (see Display 3.9). This dependence can arise through under-identification of the model itself, or can arise through empirical under-identification (due to the data and not, necessarily, to the model *per se*). In either case the results should be examined with caution.

Looking at the parameter estimates obtained through fitting the unidentified model, we can see that the variances for the error terms have not changed from those arising from fitting the identified model (compare Display 3.10 and Display 2.12). The factor variance has changed, however, as have the factor loadings. As we would expect from the discussion of under-identification in Chapter 1, the ratios of the factor loadings are the same in both outputs. The communalties (reliabilities) are also the same for both models.

Although some of the output for fitting an under-identified model can be relied on, the user should be very wary of fitting models that produce the above condition code. Every attempt should be made to discover the cause of the problem, whether it lies in the data or in the specification of the model. 'Linear dependence among parameters is a potentially serious problem, because the solution probably cannot be fully trusted' (Bentler, 1989: p. 88).

Exercise 3.5

Using the EQS program in Display 3.5, produce an under-identified model by changing the first line of the /VARIANCES paragraph to

```
F1=1; F2=1*;
```

Interpret the results of fitting this model.

Exercise 3.6

Starting with the EQS run in Display 3.6, produce an under-identified model by changing the second line of the /CONSTRAINTS paragraph to:

```
(V1,F1)=(V2,F1);
(V3,F2)=(V4,F2);
```

Interpret your results.

3.6 FITTING MODELS WITH INTERCEPTS AND MEANS

This section will briefly introduce the EQS facility simultaneously to fit models to means and variation/covariation about these means

```
EQS, A STRUCTURAL EQUATION PROGRAM       BMDP STATISTICAL SOFTWARE INC.
COPYRIGHT BY P. M. BENTLER               VERSION 3.0a (C) 1985, 1986, 1989.

  PROGRAM CONTROL INFORMATION (abbreviated)
  17  /EQUATIONS
  18    V1=1*F1 +E1;
  19    V2=1*F1 +E2;
  20    V3=1*F1 +E3;
  21    V4=1*F1 +E4;
  22  /VARIANCES
  23    F1=3.0*;
  24    E1 TO E4=0.1*;

TITLE: Single Factor Model                                          07/30/92     PAGE: 2
COVARIANCE MATRIX TO BE ANALYZED: 4 VARIABLES (SELECTED FROM 4 VARIABLES), BASED ON 15 CASES.

              RULER    GRAHAM    BRIAN    ANDREW
              V1       V2        V3       V4
RULER    V1   3.878
GRAHAM   V2   2.811    2.121
BRIAN    V3   3.148    2.267     2.655
ANDREW   V4   3.506    2.569     2.838    3.235

BENTLER-WEEKS STRUCTURAL REPRESENTATION:
    NUMBER OF DEPENDENT VARIABLES=4
        DEPENDENT V'S:  1  2  3  4
    NUMBER OF INDEPENDENT VARIABLES=5
        INDEPENDENT F'S:  1
```

```
3RD STAGE OF COMPUTATION REQUIRED 1232 WORDS OF MEMORY, PROGRAM ALLOCATE 36712 WORDS
DETERMINANT OF INPUT MATRIX IS   .17450D-02

TITLE: Single Factor Model                                        07/30/92    PAGE: 3
MAXIMUM LIKELIHOOD SOLUTION (NORMAL DISTRIBUTION THEORY)

PARAMETER       CONDITION CODE
  V4,F1         LINEARLY DEPENDENT ON OTHER PARAMETERS

TITLE: Single Factor Model                                        07/30/92    PAGE: 6
MAXIMUM LIKELIHOOD SOLUTION (NORMAL DISTRIBUTION THEORY)
*** WARNING *** TEST RESULTS MAY NOT BE APPROPRIATE DUE TO CONDITION CODE

GOODNESS OF FIT SUMMARY
INDEPENDENCE MODEL CHI-SQUARE=148.520, BASED ON 6 DEGREES OF FREEDOM
INDEPENDENCE AIC =   136.52010   INDEPENDENCE CAIC =   126.27180
       MODEL AIC =      .34792          MODEL CAIC =    -1.36013

CHI-SQUARE=2.348 BASED ON 1 DEGREES OF FREEDOM
PROBABILITY VALUE FOR THE CHI-SQUARE STATISTIC IS .12545
THE NORMAL THEORY RLS CHI-SQUARE FOR THIS ML SOLUTION IS 2.538.

BENTLER-BONETT NORMED      FIT INDEX=    .984
BENTLER-BONETT NONNORMED FIT INDEX=    .943
COMPARATIVE FIT INDEX              =    .991
```

DISPLAY 3.9 Selected output for unidentified model.

```
TITLE: Single Factor Model                                        07/30/92    PAGE: 7
MAXIMUM LIKELIHOOD SOLUTION (NORMAL DISTRIBUTION THEORY)
MEASUREMENT EQUATIONS WITH STANDARD ERRORS AND TEST STATISTICS

RULER    =V1  =        1.140*F1  +  1.000 E1
                        .043
                      26.242
GRAHAM   =V2  =         .829*F1  +  1.000 E2
                        .052
                      15.935
BRIAN    =V3  =         .927*F1  +  1.000 E3
                        .059
                      15.724
ANDREW   =V4  =        1.034*F1  +  1.000 E4
                        .000
                35901545.669

TITLE: Single Factor Model                                        07/30/92    PAGE: 8
MAXIMUM LIKELIHOOD SOLUTION (NORMAL DISTRIBUTION THEORY)
```

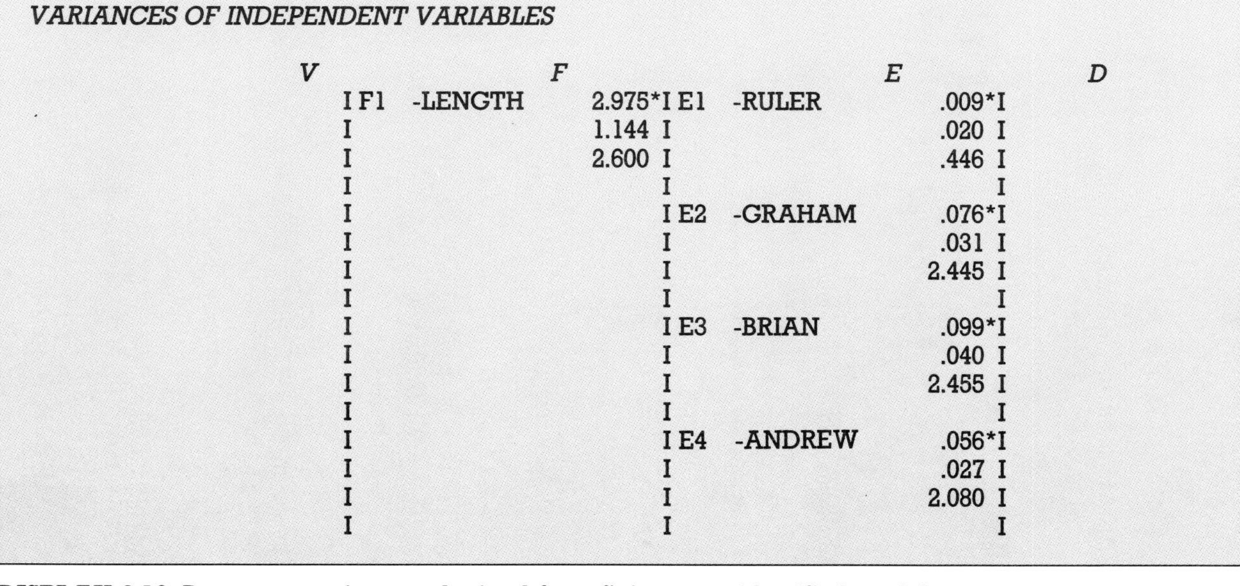

```
VARIANCES OF INDEPENDENT VARIABLES

      V                 F                          E                D
      I F1   -LENGTH       2.975*I E1   -RULER          .009*I
      I                    1.144 I                       .020 I
      I                    2.600 I                       .446 I
      I                          I                            I
      I                          I E2   -GRAHAM         .076*I
      I                          I                       .031 I
      I                          I                      2.445 I
      I                          I                            I
      I                          I E3   -BRIAN          .099*I
      I                          I                       .040 I
      I                          I                      2.455 I
      I                          I                            I
      I                          I E4   -ANDREW         .056*I
      I                          I                       .027 I
      I                          I                      2.080 I
      I                          I                            I
```

DISPLAY 3.10 Parameter estimates obtained from fitting an unidentified model.

(more details will be provided in Section 7.6). One of the simplest examples is fitting a bivariate regression model containing an intercept term (Display 2.1 illustrates an EQS run in which the intercept term has been ignored). Returning to the regression of Graham's guesses of string length on the measurements provided by a ruler (Section 1.3), the regression equation is:

$$G = a + bT + E$$

where, as before, G is Graham's guess and T the truth (provided by the ruler). E is the measurement error associated with Graham's guess. The parameters of the model, a and b, are the intercept and slope, respectively. Now, implicit in this regression model are the two additional equalities:

$$\begin{aligned} \text{Mean}(G) &= a + b\text{Mean}(T) \\ \text{Var}(G) &= b^2\text{Var}(T) + \text{Var}(E) \end{aligned}$$

There are, therefore, three more parameters to be estimated, that is, Mean(T), Var(T) and Var(E), giving five in all. In EQS we can explicitly ask for these five parameters to be estimated through the use of the following /EQUATIONS paragraph:

```
/EQUATIONS
V1=4*V999+E1;
V2=0.1*V999+1*V1+E2;
```

where V1 is measurement provided by the ruler, V2 is Graham's guess, and the new 'variable', V999, is an EQS system variable that has a constant value of 1 (it is an independent variable with zero variance). The parameters to be estimated in the first of these two equations are the mean of T (the regression coefficient for V1 on V999) and the variance of E1 (equivalent, of course, to the variance of T). The parameters to be estimated in the second equation are the intercept term, a (the regression coefficient for V2 on V999), the slope of the calibration curve, b (the regression coefficient for V2 on V1) and the variance of E2 (the variance of the errors in Graham's guesses). The choice of starting values arises from the preliminary results given in Section 1.3; their actual values, however, are not critical here.

The required EQS input file is provided in full in Display 3.11. The only other unfamiliar line in this program is the ANALYSIS=MOMENT; statement in the /SPECIFICATIONS paragraph. This statement tells EQS to analyse first moments

```
/TITLE
 Simple Regression Model With Intercept Term
 Data are guesses of string lengths
 From Display 1.1
 Graham's Guess Regressed on the Ruler
/SPECIFICATION
 CASES=15;
 VARIABLES=4;
 METHOD=ML;
 MATRIX=CORRELATION;
 ANALYSIS=MOMENT;
/LABELS
 V1=RULER;
 V2=GRAHAM;
 V3=BRIAN;
 V4=ANDREW;
 F1=LENGTH;
/EQUATIONS
 V1=4*V999+E1;              ! Except for a constant (i.e. the mean), V1=E1
 V2=0.1*V999+1*V1+E2;       ! 0.1 is the starting value for the intercept term
/VARIANCES
 E1=4*; E2=0.1*;
/MATRIX
 1.000
 0.9802  1.000
 0.9811  0.9553  1.000
 0.9899  0.9807  0.9684  1.000
/STANDARD DEVIATIONS
 1.9692  1.4563  1.6294  1.7987
/MEANS
 4.2933  3.4333  3.4267  3.7667
/END
```

DISPLAY 3.11 Regression model with an intercept term.

(means) in addition to the variances and covariances. Whenever ANALYSIS=MOMENT; is specified, EQS requires the constant V999 to be included in the /EQUATIONS paragraph below it. Selected parts of the corresponding output file are given in Display 3.12. The reader should have no difficulty in interpreting this output, particularly if reference is made to the parameter estimates already provided in Section 1.3.

3.7 SUMMARY

In this chapter we have provided a rudimentary description of model-fitting procedures based on fitting a sequence of models and

```
MEASUREMENT EQUATIONS WITH STANDARD ERRORS AND
TEST STATISTICS

RULER    =V1  =   4.293*V999  +   1.000 E1
                   .526
                  8.158
GRAHAM   =V2  =    .725*V1    +    .321*V999  +   1.000 E2
                   .039             .185
                 18.522            1.735

VARIANCES OF INDEPENDENT VARIABLES
   V                    F                    E                    D
      I                  I E1  -RULER       3.878*I
      I                  I                  1.466 I
      I                  I                  2.646 I
      I                  I                        I
      I                  I E2  -GRAHAM       .083*I
      I                  I                   .031 I
      I                  I                  2.646 I
      I                  I                        I
```

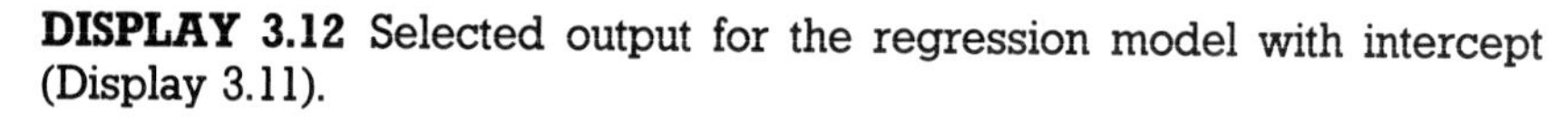

DISPLAY 3.12 Selected output for the regression model with intercept (Display 3.11).

assessing the change in fit using chi-squared statistics. We have also illustrated how different types of model can provide an equally satisfactory description of a particular data set. Finally, we have discussed two commonly occurring warnings arising from (a) parameter estimates which are out of their permitted range (or, at least, on the boundary between their permissible and impermissible values), and (b) under-identification.

Readers should now be in a position to explore more challenging data sets through the use of EQS. They should not lose sight of the fact, however, that the simpler, more straightforward models are often the most convincing!

4 Confirmatory factor analysis models

4.1 INTRODUCTION

In this chapter the fitting of some more complex models is considered. Again, each of the models leads to a prediction about the covariances or correlations between the observed variables, and model parameters are estimated by choosing values which make the observed and predicted correlations or covariances as 'close' to each other as possible, using one or other of the measures of closeness discussed in Chapter 3. Of primary interest in this chapter will be **confirmatory factor analysis** models, some simple examples of which have already arisen in Chapters 2 and 3. With this type of model, the investigator postulates that the covariances (correlations) between the observed variables arise from their relationships to a small number of underlying latent variables, more often known in this context as **factors**. With a confirmatory factor analysis model *particular* observed variables will be assumed to be **indicators** or, equivalently, will load on *particular* factors, in contrast to the **exploratory factor analysis** approach (see Everitt and Dunn, 1991), where *all* observed variables are considered to load on *all* factors.

The aim of fitting such models is generally twofold:

(a) to test some particular theory; and
(b) to produce a parsimonious description of the data.

4.2 CONFIRMATORY FACTOR ANALYSIS MODELS FOR DRUG ABUSE

Having considered some fairly simple confirmatory factor analysis models in earlier chapters, it is now time to look at a more challenging example taken from the work of Huba, Wingard and

Bentler (1981). This involves data collected from 1634 students in the seventh to ninth grades in eleven schools in the greater metropolitan area of Los Angeles. Each participant in the study completed a questionnaire about the number of times particular substances had ever been used. Responses were recorded on a five-point scale:

1. Never tried
2. Only once
3. A few times
4. Many times
5. Regularly.

Frequency-of-use data were collected for

1. Cigarettes
2. Beer
3. Wine
4. Liquor
5. Cocaine
6. Tranquilizers
7. Drugstore medication used to get 'high'
8. Heroin
9. Marijuana
10. Hashish
11. Inhalants (glue, gasoline)
12. Hallucinogenics (LSD, mescaline)
13. Amphetamine stimulants.

Product-moment correlations between the usage rates for the thirteen substances are shown in Display 4.1.

A three-factor model was postulated to explain the observed correlations, the factors being as follows:

F1: *Alcohol use*, with non-zero loadings on beer, wine, liquor and cigarettes.

F2: *Cannabis use*, with non-zero loadings on marijuana, hashish, cigarettes and wine. The cigarette variable is assumed to load on both the first and second latent variables because it sometimes occurs with both alcohol and marijuana use and other times not. The non-zero loading on wine was allowed because of the reports that wine is frequently used with marijuana and that, consequently, some of the use of wine may be an indicator of tendencies towards using cannabis.

	1	2	3	4	5	6	7	8	9	10	11	12	13
1	1.00												
2	0.447	1.000											
3	0.422	0.619	1.000										
4	0.436	0.604	0.583	1.000									
5	0.114	0.068	0.053	0.115	1.000								
6	0.203	0.146	0.139	0.258	0.349	1.000							
7	0.091	0.103	0.110	0.122	0.209	0.221	1.000						
8	0.082	0.063	0.066	0.097	0.321	0.355	0.201	1.000					
9	0.513	0.445	0.365	0.482	0.186	0.316	0.150	0.154	1.000				
10	0.304	0.318	0.240	0.368	0.303	0.377	0.163	0.219	0.534	1.000			
11	0.245	0.203	0.183	0.255	0.272	0.323	0.310	0.288	0.301	0.302	1.000		
12	0.101	0.088	0.074	0.139	0.279	0.367	0.232	0.320	0.204	0.368	0.340	1.000	
13	0.245	0.199	0.184	0.293	0.278	0.545	0.232	0.314	0.394	0.467	0.392	0.511	1.000

DISPLAY 4.1 Correlations between drug-use rates.

		F1	F2	F3
1.	Cigarettes	X	X	0
2.	Beer	X	0	0
3.	Wine	X	X	0
4.	Liquor	X	0	X
5.	Cocaine	0	0	X
6.	Tranquillizers	0	0	X
7.	Drugstore medication	0	0	X
8.	Heroin	0	0	X
9.	Marijuana	0	X	0
10.	Hashish	0	X	X
11.	Inhalants	0	0	X
12.	Hallucinogenics	0	0	X
13.	Amphetamines	0	0	X

X denotes free parameter to be estimated.

A zero denotes a parameter whose value is fixed at 0.

In addition, in this initial model, correlations between F1, F2 and F3 are fixed at zero and error terms are also not allowed to correlate.

DISPLAY 4.2 Structure of initial confirmatory factor analysis model considered for drug-use data.

F3: *Hard drug use*, with non-zero loadings for amphetamines, tranquillizers, hallucinogenics, hashish, cocaine, heroin, drug-store medication, inhalants and liquor. The use of each of these substances was considered to suggest a strong commitment to the notion of psychoactive substance use.

The structure of the proposed model is shown in Display 4.2, and the corresponding path diagram appears in Display 4.3. This path diagram is translated into a suitable EQS program in Display 4.4. When this program is run, EQS issues the following warning message

E9,E9 CONSTRAINED AT LOWER BOUND

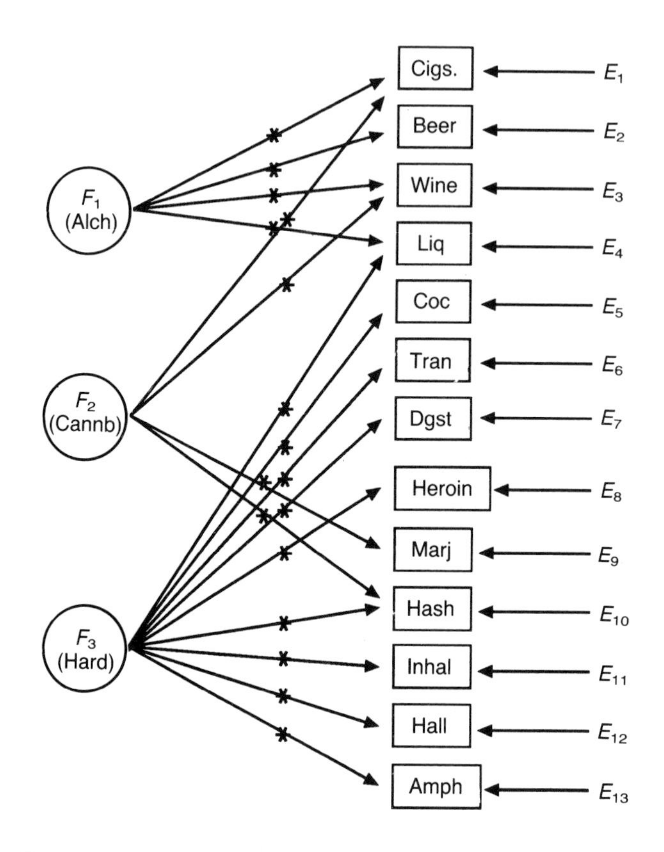

* = Free parameter to be estimated. In addition, variances of E1 to E13 to be estimated, (Note: no paths between F1, F2 and F3.)

DISPLAY 4.3 Confirmatory factor model for drug-use data: path diagram.

```
/TITLE
 Analysis of drug usage data from Huba et al.
 Model 1
/SPECIFICATIONS
 CASES=1634; VARIABLES=13; ME=ML; MA=CORR;
 AN=CORR;
/LABELS
 V1=CIGS; V2=BEER;
 V3=WINE; V4=LIQUOR;
 V5=COCAINE; V6=TRANQ;
 V7=DRUGST; V8=HEROIN;
 V9=MARIJA; V10=HASH;
 V11=INHALTS; V12=HALLUCN;
 V13=AMPHET;
/EQUATIONS
 V1=1.0*F1+1.0*F2+E1;
 V2=1.0*F1+E2;
 V3=1.0*F1+1.0*F2+E3;
 V4=1.0*F1+1.0*F3+E4;
 V5=1.0*F3+E5;
 V6=1.0*F3+E6;
 V7=1.0*F3+E7;
 V8=1.0*F3+E8;
 V9=1.0*F2+E9;
 V10=1.0*F2+1.0*F3+E10;
 V11=1.0*F3+E11;
 V12=1.0*F3+E12;
 V13=1.0*F3+E13;
/VARIANCES
 E1 to E13 = 0.5*;
 F1 to F3 = 1.0;
/MATRIX
1.000
0.447 1.000
0.422 0.619 1.000
0.436 0.604 0.583 1.000
0.114 0.068 0.053 0.115 1.000
0.203 0.146 0.139 0.258 0.349 1.000
0.091 0.103 0.110 0.122 0.209 0.221 1.000
0.082 0.063 0.066 0.097 0.321 0.355 0.201 1.000
0.513 0.445 0.365 0.482 0.186 0.315 0.150 0.154 1.000
0.304 0.318 0.240 0.368 0.303 0.377 0.163 0.219 0.534 1.000
0.245 0.203 0.183 0.255 0.272 0.323 0.310 0.288 0.301 0.302 1.000
0.101 0.088 0.074 0.139 0.279 0.367 0.232 0.320 0.204 0.368 0.340 1.000
0.245 0.199 0.184 0.293 0.278 0.545 0.232 0.314 0.394 0.467 0.392 0.511 1.000
/END
```

DISPLAY 4.4 EQS program for the model in Display 4.3.

The /PRINT paragraph controls a variety of printed information that can help to make sense of a model and the quality of the estimates. For example, the predicted covariance and/or correlation matrix can be obtained by including such a paragraph with the instructions.

```
/PRINT
COVARIANCE = YES; CORRELATION = YES;
```

The large-sample estimated correlations between parameter estimates can be found from

```
/PRINT
PARAMETER = YES;
```

Other uses of this paragraph will be described in later chapters.

DISPLAY 4.5 The /PRINT paragraph.

A message of this sort was introduced in Section 3.5 and is worrying since it often indicates a fundamental problem with the model under consideration. In such cases it is worthwhile obtaining some more detailed diagnostic information from EQS. One thing which is frequently useful is to look at the correlations between the parameter estimates. These can be obtained by including the paragraph shown in Display 4.5, in the EQS program given in Display 4.4. These estimated correlations can be useful for indicating when there are problems with a particular model. Specifically, values which lie outside the range -1 to $+1$ generally indicate an identification difficulty.

Running the program modified as in Display 4.5 to request correlations between parameter estimates, and examining the parameter correlations shows that those between the loadings of substances 9 and 10 on factor 2 and the variance of E9 are both outside the permitted range. (Note that both are flagged in the EQS output.) It is clear that the current model is not acceptable for these data. A little thought about the implications of this model for the correlations between the observed variables should serve to explain why it is not. Consider, for example, the correlation between the usage rates of liquor and marijuana. The observed value of this correlation is 0.482. In the current model however, the predicted value for this correlation is zero. The structure of the model does not allow for a non-zero correlation for these two substances and the same is true for several other pairs of substances which have substantial observed correlations. The structure

	1	2	3	4	5	6	7	8	9	10	11	12	13
1	–												
2	X	–											
3	X	X	–										
4	X	X	X	–									
5	0	0	0	X	–								
6	0	0	0	X	X	–							
7	0	0	0	X	X	X	–						
8	0	0	0	X	X	X	X	–					
9	X	0	X	0	0	0	0	0	–				
10	X	0	X	X	X	X	X	X	X	–			
11	0	0	0	X	X	X	X	X	X	X	–		
12	0	0	0	X	X	X	X	X	X	X	X	–	
13	0	0	0	X	X	X	X	X	X	X	X	X	–

'X' indicates that model can give a non-zero value for this correlation. For the model given by the path diagram in Display 4.3 and EQS program in Display 4.4, all other correlations are predicted to be zero.

DISPLAY 4.6 Form of correlation matrix predicted by model 1 for drug-use data.

of the predicted correlations for the model is shown in Display 4.6. Clearly the current model cannot give satisfactory predictions of the observed correlations even if there is very close correspondence in the places where the model allows non-zero values.

So what can be done? Obviously the model needs to be modified in some way. Such model modifications are, ideally, based on theoretical arguments, but failing these, EQS provides procedures that may help to suggest sensible changes to an existing model. (The tests may also suggest modifications that are *not* very sensible!). One such test, the **Lagrange multiplier** (LM) test, evaluates the statistical necessity of the restrictions imposed by a model, based on calculations that can be obtained on the restricted model alone. The test is very useful for evaluating whether, from a statistical point of view at least, the model could be improved substantially by freeing a previously fixed parameter. For each currently fixed parameter the test will give an estimate of the value that the parameter might take if it were to be freely estimated rather than constrained. The test also gives the change in the chi-

The phrase /LMTEST in the input file provides a default test. The default provides information on univariate tests for specific parameters fixed at zero, and for all parameters fixed at a non-zero value. It also provides for a foward stepwise procedure that, at any stage, selects, as the next parameter to be added to the multivariate test, that single fixed parameter that provides the largest contribution to the increment in the current multivariate chi-squared statistic.

Results of using /LMTEST on initial model for drug usage data

Ordered univariate test statistics

Parameter	Chi-squared	Probability	Parameter change
F2, F1	392.181	0.000	0.568
F3, F2	281.740	0.000	0.464
V9, F1	257.940	0.000	0.409
V9, F3	207.992	0.000	0.365
V2, F2	93.597	0.000	0.194
F3, F1	80.089	0.000	0.281
V4, F2	69.565	0.000	0.100
etc.			

Cumulative Multivariate Statistics

				Univariate Increment	
Param.	Chi-sq.	d.f.	Prob.	Chi-sq.	Prob.
F2, F1	392.181	1	0.000	392.181	0.000
F3, F2	659.673	2	0.000	267.492	0.000
F3, F1	739.762	3	0.000	80.089	0.000
V12, F1	760.657	4	0.000	20.895	0.000
V13, F2	777.037	5	0.000	16.380	0.000
V11, F2	793.073	6	0.000	16.036	0.000
V11, F1	808.288	7	0.000	15.215	0.000
V6, F2	820.493	8	0.000	12.205	0.000
V8, F1	831.016	9	0.000	10.522	0.001
V10, F1	834.920	10	0.000	3.904	0.048

This suggests that allowing F1 and F2 to correlate will improve the fit to the extent of decreasing the value of the chi-squared statistic by 392. The size of the correlation will be approximately 0.57. In addition, freeing the correlations between F3 and F2 and between F3 and F1, will lead to further decreases in the chi-squared statistic of about 267 and 80, respectively. All these represent very substantial improvements and the clear message is – free the correlations between factors!

DISPLAY 4.7 The /LMTEST paragraph in EQS.

squared test statistic that would result from freeing each fixed parameter. Display 4.7 illustrates how the LM test can be invoked and summarizes the results of applying the test to the current model for the drug-usage data.

It is clear that allowing the three factors to have non-zero correlations will reduce considerably the value of the chi-squared statistic and give a substantial improvement in fit. (Allowing other parameters to have non-zero values might also improve the fit, but, as a first modification of the model at least, it seems more natural to free the factor correlations.) By adding the paragraph shown in Display 4.8 to the original EQS program, the new model, whose path diagram appears in Display 4.9, can be fitted. Note that for this model, no problems with parameters constrained at lower bounds are encountered, and if correlations between parameters are examined, all lie in the permitted range. Parameter estimates for this model are shown in Display 4.10.

Before looking at the difficult but important problem of assessing the detailed fit of a model, and of comparing the fit of a number of competing models, it will be helpful to consider one further possible model for the drug-usage data. This new model allows non-zero correlations between a number of the error terms. The rationale for this model given by the original investigators is as follows:

1. Amphetamines and cocaine share a subjective 'rush' effect, so the respective error terms, E5 and E13, might be correlated.
2. Tranquillizers and heroin show a sedative effect, so that a correlation between E6 and E8 should be allowed.
3. Both tranquillizers and amphetamines are used in pill form by young adolescents, and there might be a preference expressed for pills regardless of their psychoactive effect. So a correlation between E6 and E13 might be postulated.

The following paragraph is needed

```
/COVARIANCES
 F1,F2 = 0.3*; F1,F3 = 0.3*; F2,F3 = 0.3*;
```

An alternative shorthand notation is

```
F1 to F3=0.3*;
```

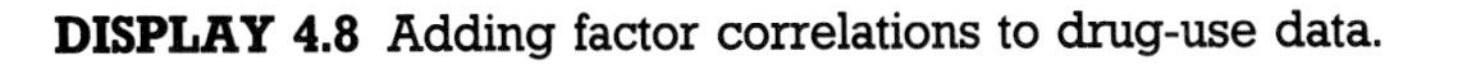

DISPLAY 4.8 Adding factor correlations to drug-use data.

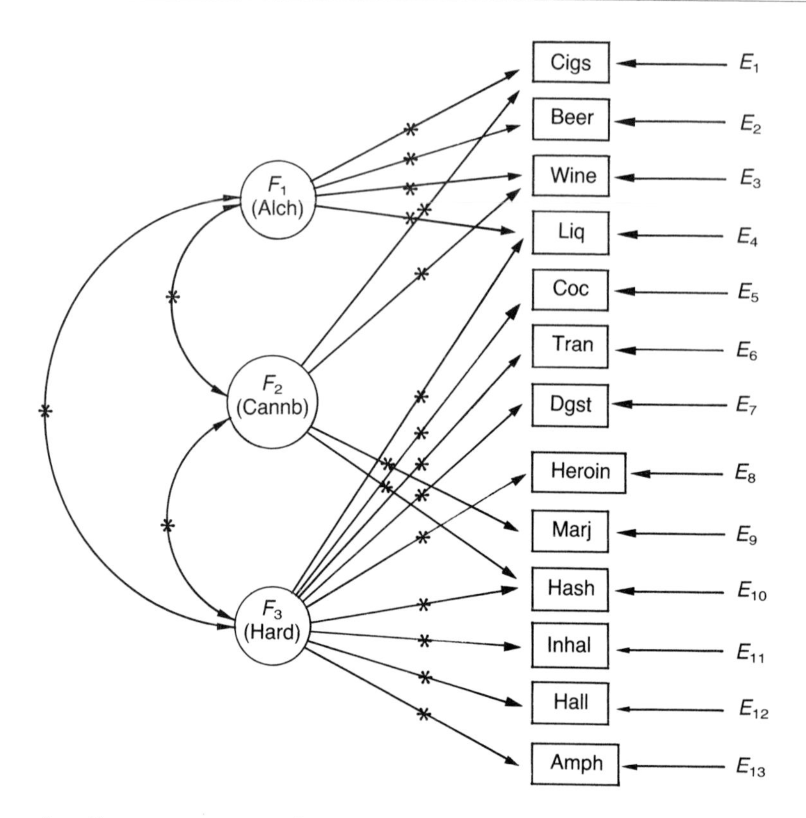

* = Free parameter to be estimated. Note paths between factors: model allows observed variables which are not indicators of factors to be correlated.

DISPLAY 4.9 Correlated factor model for drug-use data.

4. Both drugstore substances and inhalants are legally obtainable and readily available for minimal cost. Consequently, it seems reasonable to allow for a possible correlation between E7 and E11.

The path diagram for the new model is shown in Display 4.11, and the additional statements needed in the current EQS program to fit the model are given in Display 4.12. The estimates of the error term correlations are shown in Display 4.13.

Three models have now been fitted to the drug-usage correlations and it becomes of interest to choose among them using the various measures of fit given by EQS. These measures are summarized in Display 4.14, and in Display 4.15 the three models for

(1) Measurement equations with standard errors and test statistics given in EQS format

```
CIGS    = V1  =  0.358*F1 +   0.332*F2 + 1.000E1
                 0.035        0.035
                10.373        9.403
BEER    = V2  =  0.792*F1 +   1.000E2
                 0.023
                35.021
WINE    = V3  =  0.876*F1 + -0.152*F2 + 1.000E3
                 0.038        0.037
                23.290       -4.157
LIQUOR  = V4  =  0.722*F1 +   0.123*F3 + 1.000E4
                 0.024        0.023
                30.671        5.441
COCAINE = V5  =  0.465*F3 +   1.000E5
                 0.026
                18.078
TRANQ   = V6  =  0.676*F3 +   1.000E6
                 0.024
                28.124
DRUGST  = V7  =  0.359*F3     1.000E7
                 0.026
                13.601
HEROIN  = V8  =  0.476*F3 +   1.000E8
                 0.026
                18.570
MARIJA  = V9  =  0.912*F2 +   1.000E9
                 0.030
                29.954
HASH    = V10 =  0.396*F2 +   0.381*F3 + 1.000E10
                 0.030        0.029
                13.374       13.052
INHALTS = V11 =  0.543*F3 +   1.000E11
                 0.025
                21.602
HALLUCN = V12 =  0.618*F3 +   1.000E12
                 0.025
                25.233
AMPHET  = V13 =  0.763*F3 +   1.000E13
                 0.023
                39.980
```

DISPLAY 4.10 Parameter estimates for correlated factor model fitted to drug-use data.

Note the very significant z-values. F1 is well determined by beer, wine and liquor but not so well by cigarettes. F2 is almost completely determined by marijuana consumption, and F3 has as its strongest indicators tranqulizers, hallucinogenics and amphetamines.

(2) Variances of error terms with standard errors and test statistics

	E1	E2	E3	E4	E5	E6	E7
Param	0.611	0.373	0.379	0.408	0.784	0.544	0.871
SE	0.024	0.020	0.024	0.019	0.029	0.029	0.032
z	25.823	18.743	16.053	21.337	26.845	23.222	27.653

	E8	E9	E10	E11	E12	E13
Param	0.773	0.168	0.547	0.705	0.618	0.418
SE	0.029	0.044	0.022	0.027	0.025	0.021
z	26.735	3.838	24.594	25.941	24.655	19.713

Note high variances of E1, E7 and E8.

(3) Estimated correlations with standard errors and test statistics

	F1,F2	F1,F3	F2,F3
Param	0.634	0.313	0.499
SE	0.027	0.029	0.027
z	23.360	10.675	18.412

Correlations all quite substantial, particularly that of F1 and F2. Values are similar to those predicted by LMTEST – see Display 4.7.

DISPLAY 4.10 *Continued*

the drug usage data are compared using the measures. The results show that the correlated factor model is an enormous improvement on the model which constrains factor correlations to be zero, although it remains a poor fit as judged by the chi-squared statistic. The model allowing correlated error terms provides an improved fit over the previous models, even though the estimated correlations are relatively small. In this case there were seemingly sound theoretical reasons for considering the correlated errors model. In general, however, users should be cautious of attempting to improve the fit of their models by simply allowing some error

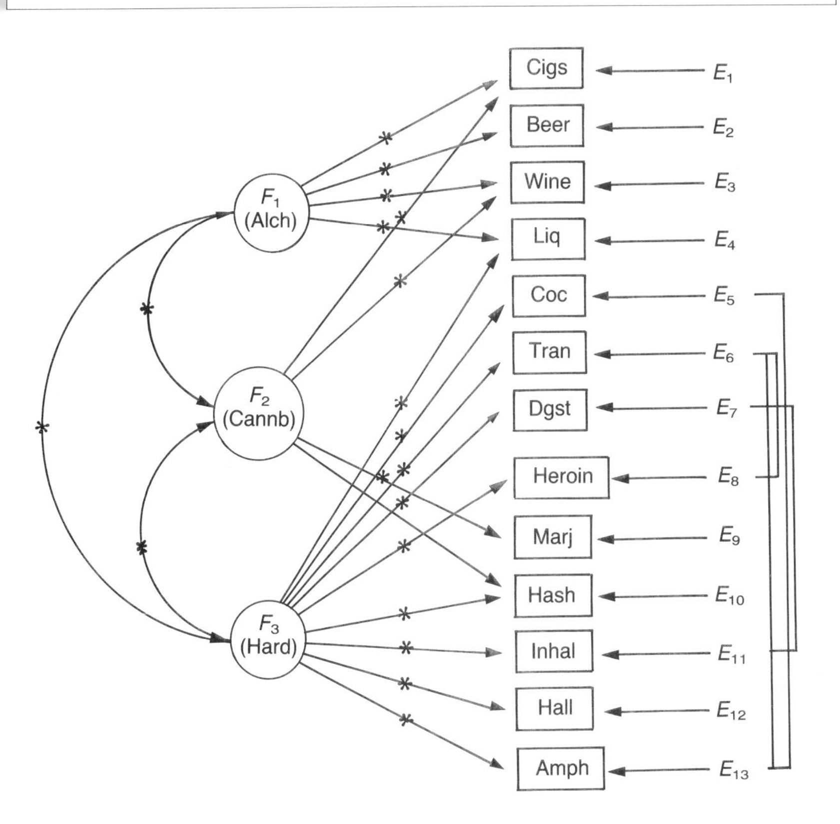

* = Free parameter to be estimated. Note extra paths between particular error terms. The covariances of these terms are now free parameters to be estimated.

DISPLAY 4.11 Correlated error model for drug-use data.

```
/COVARIANCES
 E5,E13=0.1*;
 E6,E8=0.1*;
 E6,E13=0.1*;
 E7,E11=0.1*;
```

DISPLAY 4.12 Modification to previous EQS program needed to allow for correlated error terms.

Param.	Estimated variances	Estimated covariance	Estimated correlation	SE	z
E13	0.391				
E5	0.728				
E13,E5		−0.130	−0.244	0.018	−7.237
E8	0.788				
E6	0.602				
E8,E6		0.069	0.100	0.019	3.610
E13	0.391				
E6	0.602				
E13,E6		0.058	0.120	0.020	2.883
E11	0.721				
E7	0.888				
E11,E7		0.133	0.166	0.022	6.188

Note that estimated correlations are all relatively small though highly significant.

DISPLAY 4.13 Estimates of correlations between particular error terms in model for drug-use data.

terms to have non-zero correlations. A theoretically justifiable model that provides a less satisfactory fit than a model adjusted on an *ad hoc* basis should be preferred.

The correlated factor/correlated error model still results in a significant value of the chi-squared statistic and it may be worth trying to find models which improve the fit more. The goodness-of-fit indices are however, all relatively large and it may be that only trivial discrepancies between observed and predicted correlations are leading to the large χ^2 value. Such discrepancies might, for all practical purposes, be unimportant.

Exercise 4.1

Starting values of 0.3 were used for each of the factor correlations in the second model fitted to the drug-usage data. Can you suggest more suitable values?

1. Chi-squared statistics

(a) Given for independence model, i.e. that all observed variables are independent, so that there is, essentially, 'nothing to explain' – useful baseline models against which other models can be evaluated for the gain in explanation they achieve.
(b) Given for model being fitted – based on assumption of normality. If the model is correct and the sample size sufficiently large, χ^2 is the likelihood ratio test statistic for testing the model against the alternative that the covariance matrix is unconstrained.

2. Akaike's information criterion (AIC) and Bozdogan's version of the statistic (CAIC).

These take into account both the statistical goodness of fit and the number of parameters that have to be estimated to achieve that degree of fit. The model that produces the minimum value of either statistic may, in the absence of other substantive criteria, be considered potentially the most useful.

3. Further indices of fit

EQS gives values of three other indices of fit: normed fit index (NFI); non-normed fit index (NNFI); and comparative fit index (CFI). Each of these is based on the value of the fitting function for the current model, and they have an upper limit of unity. The NNFI has the advantage of reflecting model fit very well at all sample sizes. Experience shows that these indices need to have values above 0.9 before the corresponding model can even be considered moderately adequate.

4. Residual covariance matrix

This gives the differences between the covariances predicted by the model and those observed. Values should be small and evenly distributed among variables if the model is a good representation of the data. Large residuals associated with specific variables indicate that the structural model explaining the variable is probably wrong.

A standardized version of this residual matrix is also printed out, the elements of which are in a correlational metric. If the model has been fitted to a correlation rather than a covariance matrix, the two residual matrices are the same. A frequency distribution of the standardized residuals is also provided. Ideally this should be symmetric and centred around zero.

Two averages are also printed:

(a) average of lower-triangular residual matrix components.
(b) Same as (a) but ignoring diagonal elements.

The size of the second average is generally the more critical.

DISPLAY 4.14 Goodness-of-fit assessment.

Independence model

Chi-squared = 6614.37, 78 d.f. AIC = 6458.37

Model 1: No correlations between factors
Model 2: Correlations between factors
Model 3: Correlations between factors and some error terms

	Model 1	Model 2	Model 3
Average resid.	0.1078	0.029	0.026
Average off-diag.	0.1217	0.033	0.030
Chi-squared	1091.44	323.96	209.86
df	61	58	54
p	<0.001	<0.001	<0.001
AIC	969.44	207.96	101.86
NFI	0.83	0.95	0.97
NNFI	0.80	0.94	0.97
CFI	0.84	0.96	0.98

1 v 2 Difference in chi-squared = 767.5, 3 d.f.
2 v 3 Difference in chi-squared = 114.1, 4 d.f.

Note the decline in value of AIC and the increase in NFI,NNFI and CFI. The global chi-squared and associated *p*-value appears to indicate that none of the models provides a very satisfactory fit, although models 2 and 3 are a huge improvement over model 1.

DISPLAY 4.15 Comparing the fit of three models fitted to the drug-use data.

Exercise 4.2

If the LM test procedure is applied to the correlated factor model what modifications to the model are suggested? Fit the new model. How does it compare with the correlated errors model?

Exercise 4.3

Use the LM test to suggest possible changes to the correlated factors/correlated errors model for the drug-usage data. Try to find a model which results in a non-significant χ^2 value. (Note that such 'data-driven' model modifications are not recommended in practice.)

4.3 SUMMARY

Confirmatory factor models are often used in the social and behavioural sciences. Such models specify a particular factor structure for a set of observed variables by postulating which of the variables are indicators of which factors. Additionally, correlations between some or all of the factors may be allowed, and, where appropriate, particular pairs of error terms may be free to have non-zero covariances. Such models are extremely easy to apply using EQS, although they are not always without their problems, as has been seen in this chapter. Specific points to remember are

1. Warnings that the variances of particular error terms have been set to zero usually indicate that there are problems with the current model.
2. Scales of the unobserved factors can be set either in the /VARIANCE section of the EQS program, or by fixing the loading of one of the observed variables on the factor. Although the solutions may look different they are not! This point was also covered in Chapter 3.
3. Assessing the fit of a model and comparing the fit of competing models requires more than an examination of the global chi-squared goodness-of-fit statistic.
4. The modelling procedure described in this chapter is, in some ways at least, quite different to that described in Chapter 3. Here the concern has involved a search for a suitable factor structure for a set of several manifest variables. Although called *confirmatory* factor analysis, it is in fact quite *exploratory*. The model-fitting strategies in Chapter 3, on the other hand, involved the testing of quite precise hypotheses in the context of a restricted range of models possible for the data.

Multitrait-multimethod and multiple indicator multiple cause models

5

5.1 INTRODUCTION

In this chapter two further models will be introduced and examples given of how they may be fitted using EQS. The first is used where each of a set of traits is measured by each of a set of methods and is known as the **multitrait-multimethod** (MTMM) model. The second model is used when a single latent variable is influenced by multiple causes and measured by multiple indicators, the so-called MIMIC model. Both these models are used widely in the social and behavioural sciences, and they can both be used to illustrate a number of EQS features not previously met.

5.2 MULTITRAIT-MULTIMETHOD MODELS

Many studies in the social and behavioural sciences are concerned with assessing the **reliability** and **validity** of test scores. According to Campbell and Fiske (1959): 'Reliability is the agreement between two efforts to measure the same trait through maximally similar methods. Validity is represented in the agreement between two methods to measure the same trait through maximally different methods.' In terms of the models discussed in previous chapters, reliability is concerned with the extent to which the observed indicators of latent variables contain measurement error, and validity relates to whether or not the indicators actually measure the abstract concept supposed.

Campbell and Fiske suggest that both reliability and validity can be studied by using a multitrait-multimethod (MTMM) approach: different traits (the latent variables) are measured in different ways, that is, by different methods. The MTMM idea can be formulated as a special type of confirmatory factor model in which trait factors and method factors appear. If the measurement of a trait is not affected by the method used in its measurement, then the observed variable will load only on the common factor for that trait and not on the common factor for the method. If, however, there is an effect of the method of measurement, then each observed variable will load on both the common factor for the trait and the factor for the particular method of measurement being used.

An example will help to clarify these points, and here an investigation described by Sullivan and Feldman (1979) will be used. Interest centred on people's evaluations of the two major political parties in the USA, referred to as *partisan evaluation*. This could be measured by using an attitude questionnaire or an interview schedule. It may, however, be suspected that a respondent's answers to such a questionnaire may be relatively casual and largely influenced by temporal and possibly even random events. Consequently, the validity of this particular measurement procedure may be in doubt and other approaches to measuring partisan evaluation considered. One possibility would be to spend two or three days observing each respondent and recording all conversational mentions of either party. From these a partisan evaluation score could be constructed. (This type of measurement procedure would, of course, by very expensive, even if the harassed respondent was to agree to it! It is unlikely to be a practical possibility in most situations, but is used here simply to illustrate the MTMM approach.) A third measurement method might be to question each respondent's friends and relatives, using them as 'informants' about the respondent's partisan evaluation.

In addition to partisan evaluation, the investigators were interested in a respondent's *political ideology* in terms of a liberal–conservative dimension and in their degree of *political involvement*. Both were measured by the three methods described above. Thus the data consist of measurements on three traits, partisan evaluation, political ideology and political involvement, by three different methods, interview with respondent, participant observation and the use of informants. The observed correlation matrix is given in Display 5.1. The off-diagonal elements in such a

Methods		Interview			Observation			Informant		
Traits:		1	2	3	1	2	3	1	2	3
	1	1.00								
Interv.	2	0.42	1.00							
	3	0.38	0.33	1.00						
	1	**0.51**	0.32	0.29	1.00					
Observ.	2	0.31	**0.45**	0.19	0.44	1.00				
	3	0.30	0.25	**0.39**	0.38	0.32	1.00			
	1	**0.51**	0.31	0.30	**0.62**	0.36	0.28	1.00		
Infor.	2	0.35	**0.48**	0.21	0.25	**0.68**	0.25	0.46	1.00	
	3	0.28	0.19	**0.39**	0.24	0.23	**0.59**	0.37	0.36	1.00

Within-method, cross-trait correlations are underlined.

Within-trait, cross-method correlations are in **bold**.

DISPLAY 5.1 Multitrait-multimethod correlation matrix.

multitrait-multimethod matrix can be classified into three groups. In the triangles adjacent to the main diagonal are **within-method**, **cross-trait** correlations. An example is the value 0.38 in the third row and first column, the correlation between partisan evaluation and political involvement both assessed by interview. In the diagonals of the square blocks in the rest of the table are the **within-trait**, **cross-method** correlations. An example is the value 0.51 in the fourth row and first column, the correlation between partisan evaluation assessed by interview and partisan evaluation assessed by observation. The remaining elements of the matrix are the **cross-trait**, **cross-method** correlations.

High cross-method, within-trait correlations are evidence of what Campbell and Fiske (1959) call **convergent validity**, the agreement of different methods of measuring the same trait. Low correlations elsewhere give evidence of **discriminant validity**, that the assumed different traits really are distinct.

A possible model for the correlations in Display 5.1 is represented by the path diagram shown in Display 5.2. Three latent variables representing the true scores on the three traits are postulated. In addition, three further latent variables representing the effects of the three methods of measurement are included.

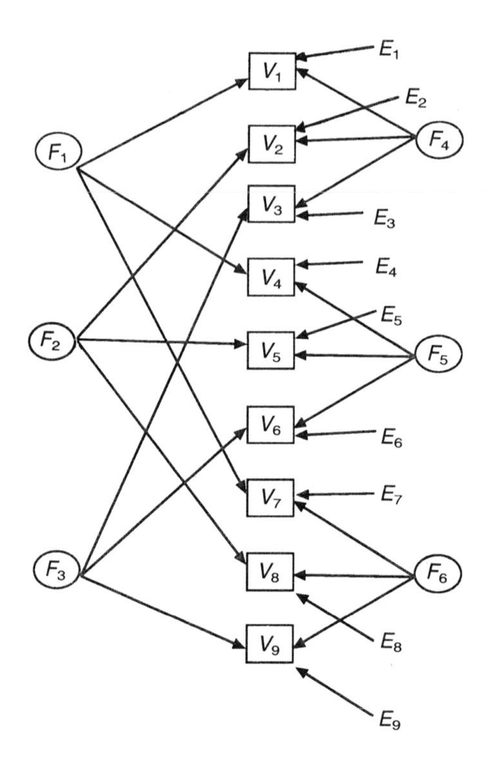

DISPLAY 5.2 Path diagram for multitrait-multimethod matrix.

Each observed measurement is assumed to be determined by a trait and a method, plus, of course, an error term. In the model specified in Display 5.2 neither trait nor method factors are assumed to be correlated. An EQS program for fitting this model is given in Display 5.3. Parameter estimates are shown in Display 5.4. Fit indices are summarized in Display 5.5.

The measurements are clearly substantially determined by the traits, although somewhat less so when the method used is interview; they are somewhat less determined by the methods, although the loadings are still moderately large. The goodness-of-fit summary in Display 5.5 indicates that the current model does not fit very well. That this would be the case could have been predicted from noting that in the observed correlation matrix there are a number of substantial correlations for variables having neither a trait nor method factor in common – the current model predicts that all such correlations will be zero.

An obvious addition to the existing model is to allow the trait

```
/TITLE
Analysis of multitrait-multimethod correlation matrix
No correlations between method or trait factors
/SPECIFICATIONS
 CASES=500; VARIABLES=9; METHOD=ML; MATRIX=CORR; ANALYSIS=
 CORR;
/LABELS
 V1=M1T1; V2=M1T2; V3=M1T3; V4=M2T1; V5=M2T2; V6=M2T3; V7=M3T1;
 V8=M3T2; V9=M3T3;
/EQUATIONS
 V1=1.0*F1+1.0*F4+E1;  ! F1=partisan evaluation trait,  F4=interview method
 V2=1.0*F2+1.0*F4+E2;  ! F2=political ideology trait
 V3=1.0*F3+1.0*F4+E3;  ! F3=political involvement trait
 V4=1.0*F1+1.0*F5+E4;  ! F5=participant observation method
 V5=1.0*F2+1.0*F5+E5;
 V6=1.0*F3+1.0*F5+E6;
 V7=1.0*F1+1.0*F6+E7;  ! F6=informant method
 V8=1.0*F2+1.0*F6+E8;
 V9=1.0*F3+1.0*F6+E9;
/VARIANCES
 F1 TO F6=1.0;         ! Scales of method and trait
                       ! Factors fixed at one
 E1 TO E9=0.3*;
/MATRIX

! Correlations here

/END
```

DISPLAY 5.3 EQS program for fitting model specified by path diagram in Display 5.2 to multitrait-multimethod correlations in Display 5.1.

factors or method factors to correlate. The two possible models are represented by the path diagrams in Displays 5.6(a) and 5.6(b). The changes needed to the previous EQS program to fit each of these models are shown in Display 5.7. Some measures of fit for each model are given in Display 5.8, and the estimated correlations appear in Display 5.9.

Although the estimated correlations are quite large for both models, the correlated method model clearly provides the better fit. Taking this as the current model, the LMTEST can be applied to try to identify possibly useful modifications. Some of the results from this test are shown in Display 5.10. The test suggests that freeing the parameters V1,F6 and V8,F1 will improve the fit of the correlated method factors model. These parameters reflect the following:

```
Measurement equations with standard errors and test statistics

M1T1 = V1 = 0.595*F1 + 0.371*F4 + 1.000E1
            0.043      0.057
            13.842     6.538
M1T2 = V2 = 0.479*F2 + 0.471*F4 + 1.000E2
            0.043      0.068
            11.058     6.957
M1T3 = V3 = 0.420*F3 + 0.442*F4 + 1.000E3
            0.047      0.067
            8.968      6.634
M2T1 = V4 = 0.667*F1 + 0.513*F5 + 1.000E4
            0.042      0.070
            16.018     7.355
M2T2 = V5 = 0.741*F2 + 0.375*F5 + 1.000E5
            0.045      0.056
            16.561     6.732
M2T3 = V6 = 0.717*F3 + 0.301*F5 + 1.000E6
            0.054      0.052
            13.193     5.741
M3T1 = V7 = 0.791*F1 + 0.375*F6 + 1.000E7
            0.043      0.052
            18.437     7.234
M3T2 = V8 = 0.809*F2 + 0.420*F6 + 1.000E8
            0.044      0.054
            18.272     7.775
M3T3 = V9 = 0.727*F3 + 0.369*F6 + 1.000E9
            0.054      0.054
            13.484     6.779

Key to variable names:
M1 = interview, M2 = participant observation, M3 = informant;
combined with
T1 = partisan evaluation, T2 = political ideology, T3 = political
involvement.
```

DISPLAY 5.4 Parameter estimates obtained from running EQS program in Display 5.3 on correlations in Display 5.1.

V1,F6: partisan evaluation measured by interview should be allowed to load on the method factor for informant.

V8,F1: political ideology measured by informant should be allowed to load on the trait factor for partisan evaluation.

(1) Chi-squared = 226.120, 18 d.f., $p < 0.001$
(2) AIC = 190.120
(3) Fit indices
NFI = 0.872
NNFI = 0.760
CFI = 0.880
(4) Average off-diagonal absolute standardized residual = 0.2151.

Note that all methods of assessing fit suggest that the current model is very poor.

DISPLAY 5.5 Goodness-of-fit summary for first model fitted to multitrait-multimethod correlations.

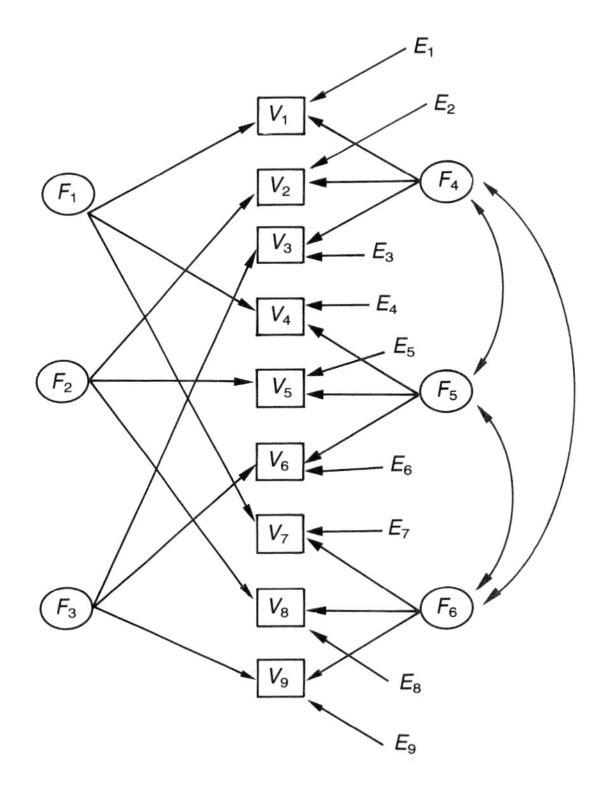

DISPLAY 5.6(a) Path diagram for multitrait-multimethod matrix (correlated method factors).

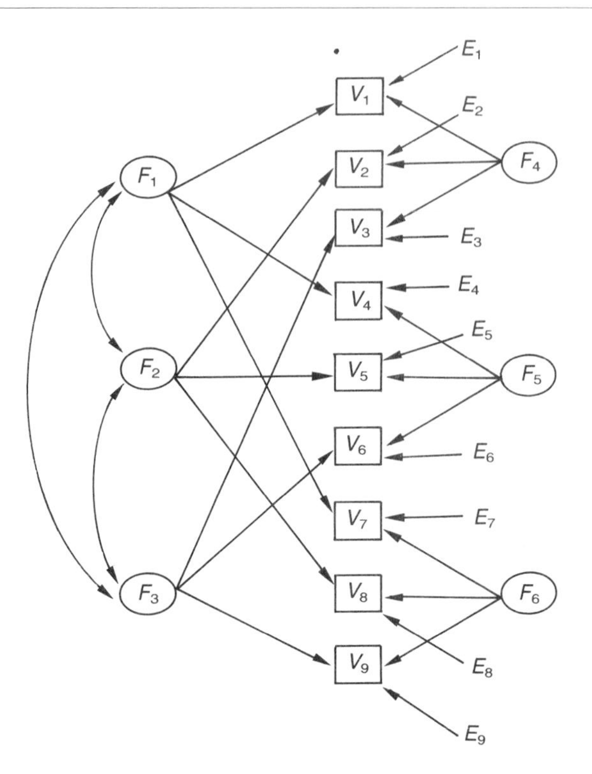

DISPLAY 5.6(b) Path diagram for multitrait-multimethod matrix (correlated trait factors).

```
(a) Trait factors correlated
  /COVARIANCES
   F1 TO F3=0.3*;
(b) Method factors correlated
  /COVARIANCES
   F4 TO F6=0.3*;
```

DISPLAY 5.7 Additional paragraphs needed to fit correlated trait factors and correlated method factors models.

Would the inclusion of such terms in the current model be sensible? Examining first V1,F6, it seems difficult to justify the inclusion of such a term. There appears to be no theoretical reason why an observed variable should load on more than the one

Measure of fit	Correlated trait factors	Correlated method factors
(1) Chi-squared	43.687	25.046
d.f.	15	15
p	<0.001	0.049
(2) AIC	13.686	−4.954
(3) Fit indices		
(a) NFI	0.975	0.986
(b) NNFI	0.960	0.986
(c) CFI	0.983	0.994

All fit measures suggest that the correlated methods model is to be preferred.

DISPLAY 5.8 Comparison of fit of correlated trait factors and correlated method factors models.

(1) Trait factor correlations

Parameter	Estimate	SE	z
F1,F2	0.518	0.041	12.578
F1,F3	0.487	0.046	10.655
F2,F3	0.407	0.048	8.405

(2) Method factor correlations

Parameter	Estimate	SE	z
F4,F5	0.730	0.043	16.891
F4,F6	0.743	0.040	18.424
F5,F6	0.685	0.038	18.085

Note that all correlations are very significantly different from zero and those between methods are substantially higher than those between traits.

DISPLAY 5.9 Estimated trait and method factor correlations.

method factor. In the case of the parameter V8,F1, however, there may be an argument for its inclusion since the variables may not be 'pure' measures of a trait. It is perhaps not too difficult to imagine that political ideology as assessed by informant is contaminated in some way by the partisan evaluation of an individual.

	Cumulative multivariate statistics			Univariate increment	
Param.	Chi-sq.	d.f.	p	Chi-sq.	p
V1,F6	8.129	1	0.004	8.129	0.004
V8,F1	17.772	2	0.000	0.644	0.002

Allowing paths between V1,F6 and V8,F1 will substantially decrease the chi-squared statistic. But are the paths both sensible?

DISPLAY 5.10 LMTEST on correlated method factors model.

Exercise 5.1.

Fit this new model and interpret your results.

A further model that might be considered is one that allows correlations between method factors *and* correlations between trait factors. A suitable EQS program is shown in Display 5.11. (Notice that in this program the method and trait factors have also been labelled.) Running this program results in a number of problems. The first is that the program stops after thirty iterations, although the iterative procedure has not converged satisfactorily. This occurs since EQS only allows this number of iterations by default. The number of iterations can be increased quite simply by including in the program the paragraph shown in Display 5.12. Here, however, increasing the number of iterations does not appear to help. The iterative procedure will not converge. A clue to the problem is provided in the EQS output where the following message appears repeatedly:

IN ITERATION = n, MATRIX W_CFUNT MAY NOT BE POSITIVE DEFINITE.

This implies that the current parameter values lead to a poorly behaved predicted covariance matrix (in technical terms, one that is not positive definite). This quite often happens when initial values are poorly chosen, but is not generally a problem unless, as appears to be the case here, the program cannot find improved estimates that will lead to convergence. In such cases, the solution is to provide new and better starting values. Here such values might be taken from one or other of the solutions already obtained.

```
/TITLE
Analysis of multimethod – multitrait data
Correlated method factors and correlated trait factors
/SPECIFICATION
 CASES=500; VARIABLES=9; METHOD=ML; MATRIX=CORR;
 ANALYSIS=CORR;
/LABELS
 V1=M1T11; V2=M1T22; V3=M1T3; V4=M2T1; V5=M2T2;
 V6=M2T3; V7=M3T1; V8=M3T2; V9=M3T3;
 F1=PEVAL; F2=PIDEO; F3=PINVOLVE;
/EQUATIONS

AS IN DISPLAY 5.3

/VARIANCES
 F1 TO F6=1.0;
 E1 TO E9=0.3*;
/COVARIANCES
 F1 TO F3=0.3*;
 F4 TO F6=0.3*;
/MATRIX

CORRELATIONS HERE

/END
```

DISPLAY 5.11 EQS program for fitting model which allows correlated method factors and correlated trait factors.

The /TECHNICAL paragraph allows the user to specify number of iterarations and the value of the convergence criterion. In addition, a number of more technical parameters can be set. To change the number of iterations from the default value of 30 requires the following instructions:

```
/TECHNICAL
 ITR=50;
```

DISPLAY 5.12 The /TECHNICAL paragraph in EQS.

Exercise 5.2.

Investigate the fitting of this model in more detail.

5.3 MULTIPLE INDICATOR MULTIPLE CAUSE (MIMIC) MODELS

The simplest case of a MIMIC model is that where a single, unobservable latent variable is influenced by multiple causes and measured by multiple indicators. To introduce this type of model, consider the following simple example, which involves an investigation of attitudes among women in less developed countries. Variables recorded were a *modernization index* (V1), a *modern objects scale* (V2) and a *media exposure index* (V3), all considered to be indicators of a latent variable referred to as *modernization orientation*. In addition, each woman's place of birth – city or rural (V4) – was noted and her educational level (V5) assessed. These two variables are assumed to affect a woman's modernization orientation. (A warning bell should ring here: place of birth is clearly not a continuous variable, let alone normally distributed. It is a **binary** variable, that is, one with only two possible values. As such its use in this type of analysis can be questioned. Here the problem will be conveniently ignored, but in Chapter 8 the topic of binary variables will be the subject of more detailed discussion.) The correlations between the five observed variables found from observations on 593 women are given in Display 5.13. Also given in this display is the path diagram of the proposed model. Here a latent variable appears as a *dependent* rather than an *independent* variable.

A suitable EQS program for fitting the model is given in Display 5.14. Note the presence of the disturbance or D term in one of the

	V1	V2	V3	V4	V5
V1	1.00				
V2	0.62	1.00			
V3	0.77	0.58	1.00		
V4	0.54	0.42	0.59	1.00	
V5	0.72	0.57	0.70	0.52	1.00

DISPLAY 5.13 Correlations from the women's attitudes data.

```
/TITLE
MIMIC model for women's attitudes data
/SPECIFICATIONS
 CASES=593; VARIABLES=5; METHOD=ML; MATRIX=CORR;
 ANALYSIS=CORR;
/LABELS
 V1=M1, V2=MOS; V3=MEI; V4=POB; V5=EDUCATION;
 F1 =MO;
/EQUATIONS
 V1=F1+E1;                ! F1 fixed to have scale of V1
 V2=1*F1+E2;
 V3=1*F1+E3;              ! V1, V2, V3 are indicators of F1
 F1 =1*V4+1*V5+D1;        ! V4 and V5 are 'causes' of F1
                          ! D1 is disturbance term
/VARIANCES
 V4=1.0; V5=1.0; E1 TO E3=0.3*;
 D1=1.0*;                 ! variance of D1 given SV of 1
/COVARIANCES
 V4,V5=0.52;              ! covariance of V4 and V5 fixed at
                            observed value
/MATRIX

CORRELATION MATRIX HERE

/END
```

DISPLAY 5.14 EQS program for fitting the model specified in Display 5.13.

equations, this being the equivalent for dependent latent variables of the E term for observed variables (see Display 2.5 and Section 3.4). Note also that the scale of the latent variable *must* in this case be fixed by setting it to that of one of the indicator variables. The scale cannot be set in a /VARIANCES paragraph since here the latent variable appears on the left-hand side of an equation as a dependent variable. Lastly, note that the variances and covariances of the observed *causal* variables are fixed at their observed values. Here they are not considered as free parameters to be estimated. What would happen if they were all allowed to be free? EQS would simply find estimates equal to the observed values and give identical results for other parameter estimates, goodness-of-fit statistics, and so on.

Parameter estimates and the standardized residual covariance matrix are shown in Display 5.15. The measures of goodness-of-fit

(1) Measurement equations with standard errors and test statistics

```
V1 = MI  = 1.000F1  + 1.000E1
V2 = MOS = 0.777*F1 + 1.000E2
           0.041
           19.146
V3 = MEI = 0.985*F1 + 1.000E3
           0.036
           27.704
```

(2) Construct equation with standard errors and test statistics

```
F1 = MO  = 0.262*V4 + 0.581*V5 + 1.000D1
           0.0268     0.029
           9.505      19.994
```

Note that education (V5) has the highest regression coefficient. It is this variable which has the greatest effect on the latent variable, modernization orientation.

(3) Standardized residual matrix

	V1	V2	V3	V4	V5
V1	0.000				
V2	0.013	0.000			
V3	0.001	−0.018	0.000		
V4	−0.024	−0.190	0.034	0.000	
V5	0.002	0.012	−0.007	0.000	0.000

Note that all residuals are relatively small, except that for V2,V4.

DISPLAY 5.15 Parameter estimates and residual correlations from running the EQS program in Display 5.14 on the correlations in Display 5.13.

(1) Chi-squared = 11.204, d.f. = 4, $p = 0.024$
(2) AIC = 3.204
(3) Fit Indices
    NFI = 0.993
    NNFI = 0.989
    CFI = 0.996

A reasonable fit is indicated.

DISPLAY 5.16 Measures of fit for MIMIC model for the women's attitudes data.

shown in Display 5.16 indicate that the model fits the data reasonably well, although there are one or two moderate residual correlations (that between place of birth and the media exposure index, for example). Use of the LMTEST shows that the addition of a direct path between these two variables will substantially improve the fit.

As a further illustration of fitting a MIMIC model using EQS, an example from criminological research described by Smith and Patterson (1984) will be used. Random samples of persons in sixty residential neighbourhoods were interviewed regarding such issues as victimization experiences, neighbourhood safety and evaluation of police performance. The data here are those obtained from a sample of people living alone. (A further, rather small, warning bell might be required here: the sample of individuals is obtained in a rather special way and this *might* present problems. Again, any such problems will be ignored here, but will be taken up in Chapter 8.) The seven variables observed were as follows:

V4: number of self-reported prior victimizations in the last twelve months;
V5: respondent's age;
V6: respondent's sex (again a binary variable, coded 1 for female and 0 for male);
V7: the rate of personal and property victimization per 100 households in the respondent's neighbourhood.

Variables V1, V2 and V3 were responses to three questions asking respondents how likely they thought it was that they would be victims of robbery (V1), burglary (V2) and vandalism (V3) during the next year. Response categories on these items ranged from 'not at all likely' to 'very likely'.

The proposed model was that variables V1, V2 and V3 acted as indicators of a latent variable, *perceived risk of victimization*, 'caused' by the last four variables. The path diagram for the model appears in Display 5.17 and the observed correlation matrix in Display 5.18.

The EQS program for this problem appears in Display 5.19. (Note the /DIAGRAM paragraph which enables EQS to print out a path diagram for the model.) Parameter estimates are given in Display 5.20 and fit indices in Display 5.21. The results indicate that prior victimization experience, and residing in a high crime neighbourhood significantly increase a respondent's perceived risk

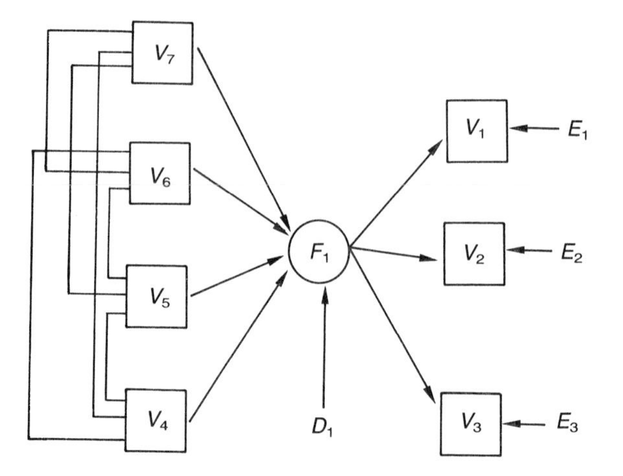

DISPLAY 5.17 Path diagram for MIMIC model on crime data.

	V1	V2	V3	V4	V5	V6	V7
V1	1.000						
V2	0.575	1.000					
V3	0.540	0.598	1.000				
V4	0.169	0.240	0.246	1.000			
V5	−0.014	−0.144	−0.128	−0.184	1.000		
V6	−0.023	−0.088	−0.092	−0.148	0.236	1.000	
V7	0.224	0.215	0.182	0.168	−0.027	−0.102	1.000

DISPLAY 5.18 Correlation matrix from Smith and Patterson (1984) data.

of victimization. Age appears to have a significant negative effect on the latent variable. According to the simple z-tests, sex of respondent has no significant effect on perceived risk, and this might perhaps be taken as justification for removing this term from the model. Unfortunately however, the simple z-tests do not take into account possible correlations between parameter estimates, and what is really needed to decide whether particular currently free parameters might be set to zero are the equivalent of multivariate tests. Such tests are provided in an EQS program by the addition of the paragraph described in Display 5.22. Using this test with the program given in Display 5.19 gives the result shown in Display 5.23. This suggests that removing the path from respondent's sex to perceived risk from the model will lead to a

```
/TITLE
Analysis of correlation matrix from Smith and Patterson
Latent variable—perceived risk of victimization
1. Victim of robbery?         ! Indicator variable
2. Victim of burglary?        ! Indicator variable
3. Victim of vandalism?       ! Indicator variable
Four exogenous variables
1. No of self-reported victimizations in last 12 months
2. Respondent's age
3. Sex
4. Personal and property   ! Note detailed title information
/SPECIFICATIONS
 CASES = 1500; VARIABLES = 7; METHOD = ML;
 MATRIX = CORR; ANALYSIS = CORR;
/LABELS
 V4 = prior_vict; V5 = Age; V6 = Sex; V7 = Vict_rate;
 V1 = prob_rob; V2 = prob_burg; V3 = prob_van;
 F1 = perc_risk;
/EQUATIONS
 F1 = 1.0*V4 + 1.0*V5 + 1.0*V6 + 1.0*V7 + D1;
 V1 = 1.0F1 + 1.0E1;
 V2 = 1.0*F1 + E2;
 V3 = 1.0*F1 + E3;
/VARIANCES
 V4 TO V7 = 1.0;
 E1 TO E3 = 0.3*;
 D1 = 0.3*;
/COVARIANCES
 V4,V5 = -0.184; V4,V6 = -0.148; V4,V7 = 0.168;
 V5,V6 = 0.236; V5,V7 = -0.027; V6,V7 = -0.102;
/MATRIX

 CORRELATIONS HERE

/DIAGRAM
 V4  V5  V6  V7;
 F1;
 D1;
 V1  V2  V3;
/END
```

DISPLAY 5.19 EQS program for fitting the model specified by the path diagram in Display 5.17 to the correlations in Display 5.18.

```
(1) Measurement equations with standard errors and test statistics

prob_rob = V1 = 1.000F1  + 1.000E1
prob_bur = V2 = 1.124*F1 + 1.000E2
                0.046
                24.278
prob_van = V3 = 1.052*F1 + 1.000E3
                0.044
                23.950

(2) Construct equations with standard errors and test statistics

perc_risk = F1 = 0.169*V4 + -0.059*V5 + -0.012*V6 + 0.163*V7 + 1.000D1
                 0.021      0.020       0.020       0.020
                 8.214      -2.933      -0.584      8.070
```

Note the non-significant z-statistic for the regression coefficient of V6. It appears that prior victimizations (V4) and victimization rate (V7) have the greatest effect on the perceived risk latent variable.

```
(3) Standardized solution

V1 = 0.714*F1 +  0.700E1
V2 = 0.802*F1 +  0.597E2
V3 = 0.751*F1 +  0.660E3
F1 = 0.236*V4 + -0.083*V5 + 0.016*V6 + 0.228*V7 + 0.925D1
```

DISPLAY 5.20 Parameter estimates from MIMIC model for Smith and Patterson (1984) data.

(1) Chi-squared = 43.597, d.f. = 8, $p < 0.001$
(2) AIC = 27.597
(3) Fit indices
  (a) NFI = 0.976
  (b) NNFI = 0.949
  (c) CFI = 0.980

Note that value of chi-squared indicates a very poor fit, although the values of the three fit indices are all relatively high.

DISPLAY 5.21 Goodness-of-fit measures for MIMIC model fitted to Smith and Patterson (1984) data.

non-significant increase in the chi-squared statistic and, consequently, will not seriously damage the fit of the model.

Since the fit of the current model appears to be relatively poor (see Display 5.24), it may be helpful to use the LMTEST

The /WTEST paragraph is designed to determine whether sets of parameters that were treated as free in the model could in fact be simultaneously set to zero without substantial loss in model fit.

DISPLAY 5.22 The /WTEST paragraph.

	Cumulative multivariate statistics			Univariate increment	
Param.	Chi-sq.	d.f.	*p*	Chi-sq.	*p*
F1,V6	0.341	1	0.559	0.341	0.559

Result indicates that there will be no significant deterioration in fit (no significant increase in chi-squared value) if the path F1,V6 is removed from the model.

DISPLAY 5.23 Results of WTEST on MIMIC model for Smith and Patterson (1984) data.

(1) Chi-squared = 43.939, d.f. = 9, $p < 0.001$
(2) AIC = 25.939
(3) Fit indices
  (a) NFI = 0.976
  (b) NNFI = 0.955
  (c) CFI = 0.981

Note that all the indicators of fit remain about the same as in Display 5.21.

DISPLAY 5.24 Fit indices for MIMIC model without path from sex to perceived risk.

procedure to suggest possible modifications. This leads to the results shown in Display 5.25. Clearly, allowing a direct path from age to likelihood of robbery will substantially improve the fit – see Display 5.26. The results from the new model are very similar to those from the initial model, with the addition of a significant positive effect of age on the robbery indicator of percieved risk. Thus, although age generally decreases the subjective likelihood of

Cumulative multivariate statistics				Univariate increment	
Param.	Chi-sq.	d.f.	p	Chi-sq.	p
V1,V5	29.273	1	0.000	29.273	0.000
V1,V4	33.353	2	0.000	4.081	0.043
V1,V7	37.321	3	0.000	3.968	0.045

Including a path from V5 to V1 should produce a substantial improvement in the fit of the model.

DISPLAY 5.25 Results from LMTEST on MIMIC model for Smith and Patterson (1984) data.

(1) New equation involving V1

V1=1.0F1+1.0*V5+1.0E1;

(2) Chi-squared=14.056, d.f.=8, p=0.08
(3) AIC = −1.94
(4) Fit indices
 (a) NFI =0.992
 (b) NNFI=0.991
 (c) CFI = 0.997

Considerable improvement in fit is indicated by all measures.

DISPLAY 5.26 Fit of MIMIC model including path from age to likelihood of robbery.

victimization, older respondents are more likely to see themselves as victims of robbery, holding constant their general perception of victimization risk. Another way to interpret the result is to say that age is positively related to the variance in perceived risk of robbery not accounted for by the underlying latent variable, subjective probability of victimization.

Exercise 5.3

Apply the LMTEST to the correlated traits model for the Sullivan and Feldman (1979) data (see Display 5.6(a)). How do the results compare to applying the same test to the correlated methods model reported in the text?

Exercise 5.4

Fit the MIMIC model specified by the path diagram in Display 5.13 to the womens' attitudes data but fix the scale of F1 to that of V2. Satisfy yourself that the result is the same as that given by the EQS program in the text where V1 is used to set the scale of F1.

5.4 SUMMARY

In this chapter two further classes of model used frequently in the social and behavioural sciences have been discussed. The first, multitrait-multimethod models, can be helpful in investigating questions of reliability and validity of test scores. These models are similar in many respects to the confirmatory factor models described in Chapter 4. Multiple indicator multiple cause models are useful in situations when an underlying latent variable having a number of observed variable indicators is thought to be influenced by other manifest variables.

Both classes of model may be fitted simply using EQS, but a number of points need to be kept firmly in mind:

1. The selection of starting values for some MTMM models may be more critical than for other confirmatory factor analysis models.
2. In MIMIC models the latent variable(s) appear on the left-hand side of the equations in the EQS program, that is, they are dependent variables. Consequently, their scales *must* be fixed by setting them to those of one or other of the observed variables. The scales *cannot* be fixed in the /VARIANCE section.

6 Models for longitudinal data

6.1 INTRODUCTION

This chapter describes a variety of models suitable for the analysis of longitudinal data. There are two main emphases. The first is the estimation and testing of causal effects. Examples have already been given of analyses that have attempted to describe causal relationships between variables all measured at a single time. Those to be described in this chapter are similar in many respects, but two features allow us to reach more confident conclusions. Firstly, where it is reasonable to rule out a causal role of later variables on earlier ones (a variety of special effects such as the expectations that subjects have about the future may prevent this), then the number of contending causal patterns that need to be considered is enormously reduced. Secondly, models are introduced that examine the causal effects between factors, rather than directly between their fallible indicator variables. Such relationships between factors give us 'cleaner' estimates of effects between variables in the sense that we have attempted to remove many (sometimes all) of the effects of measurement error. An example is presented that shows how the failure to account for measurement error can result in misleading conclusions being reached about the causal relationships that might be present in data.

The second emphasis of the chapter is on alternative basic models suitable for modelling the evolution over time of measurements made on various response variables. This emphasis is of particular importance to those involved with developmental processes and draws on McArdle and Aber (1990). We begin with models in which continuity over time in the amount of some characteristic or attribute is accounted for by allowing the amount at each occasion to have a direct impact on the amount at subsequent occasions, though without specifying exactly how this comes about. The performance of these models, which are often referred to as **autoregressive** models, is then compared to that of two other forms of model. In the first, any and all previously acquired characteristic is carried forward to the next occasion and is then added to or partially used up. The random element of the possible additions or subtractions at each occasion gives rise to the name **random walk** model. In the second, the amount of the characteristic or attribute follows a trajectory over time, with both the initial amount and the rate of increase or decrease varying from subject to subject. Such a model structure is referred to as a **growth curve** model.

These various models are initially explained and described for processes in which just a single characteristic is being examined. All these models may also be used in situations in which several characteristics that may be influencing each other are being examined. We give one example of such a circumstance, illustrating how to resolve whether one characteristic is influencing a second one, or the second is influencing the first.

6.2 ABILITY SCORES OVER TIME

Display 6.1 gives, for a group of children, the means, standard deviations and correlations for four measurements of ability (percentage of test questions correct) made at the ages of 6, 7, 9 and 11 (Osbourne and Suddick, 1972). As one would expect, the means increase progressively over time, as do the standard deviations. Such data are more often presented as standardized IQ scores rather than the 'raw' scores given here, and would then show

	Age 6	Age 7	Age 9	Age 11
Mean	18.034	25.819	35.255	46.593
Standard deviation	6.374	7.319	7.796	10.386
Correlation matrix:				
	1.000			
Age 7	0.809	1.000		
Age 9	0.806	0.850	1.000	
Age 11	0.765	0.831	0.867	1.000

DISPLAY 6.1 Data on ability: percentage of test answers correct at ages 6, 7, 9 and 11 (Osbourne and Suddick, 1972).

little variation over time in either means or standard deviations. However, the correlation matrix of scores over time would continue to look as this one does, with higher correlations between adjacent measures than between those further apart in time.

6.3 DIRECT AND INDIRECT EFFECTS

Perhaps the simplest plausible model that we could attempt to fit to these data is that with the structure shown in Display 6.2. In that model, ability at age 6 directly influences ability at age 7, ability at age 7 directly influences ability at 9, and so on. What this model rules out is any direct links between non-adjacent measures, namely those from age 6 to ages 9 and 11, and that from age 7 to age 11 (the dotted lines). Thus, there is no explicit allowance for 'sleeper effects' or the showing of some early ability that then lies dormant before being reawakened some years later. All long-term continuity must occur indirectly through the intervening measures. Though 'sleeper effects' may have some theoretical appeal in particular circumstances, often the biological or social mechanisms that are required to 'carry' such hidden effects over time are rather convoluted. It is therefore natural to begin with models that exclude them. Since this model allows the previous value of a variable to influence the current value, but no earlier values to do so, the model is often referred to as a **first-order autoregressive** model (and also as a **first-order Markov** model).

For this simple structure, the expected correlation matrix is readily calculated using simple path tracing rules and is shown in Display 6.3. The individual *r*s are just the correlation coefficients between adjacent measures and so the correlation between age 6 and 9 is the product of the correlations between the measures at 6

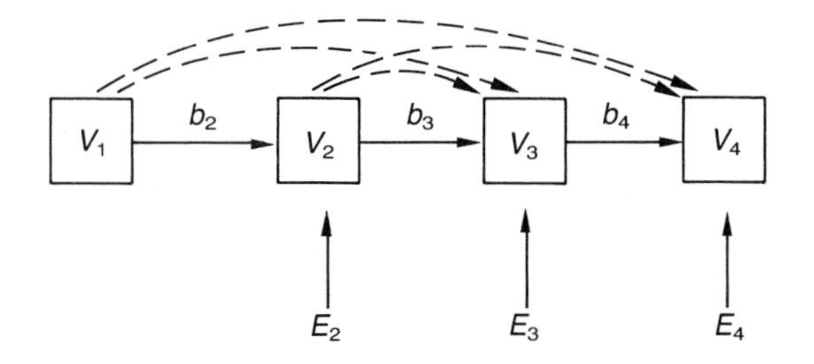

DISPLAY 6.2 A simple model of continuity in ability scores.

	V1	V2	V3	V4
V1	1			
V2	b_2	1		
V3	b_2b_3	b_3	1	
V4	$b_2b_3b_4$	b_3b_4	b_4	1

DISPLAY 6.3 Expected correlation matrix from simple first-order autoregressive model for the test data in Display 6.1.

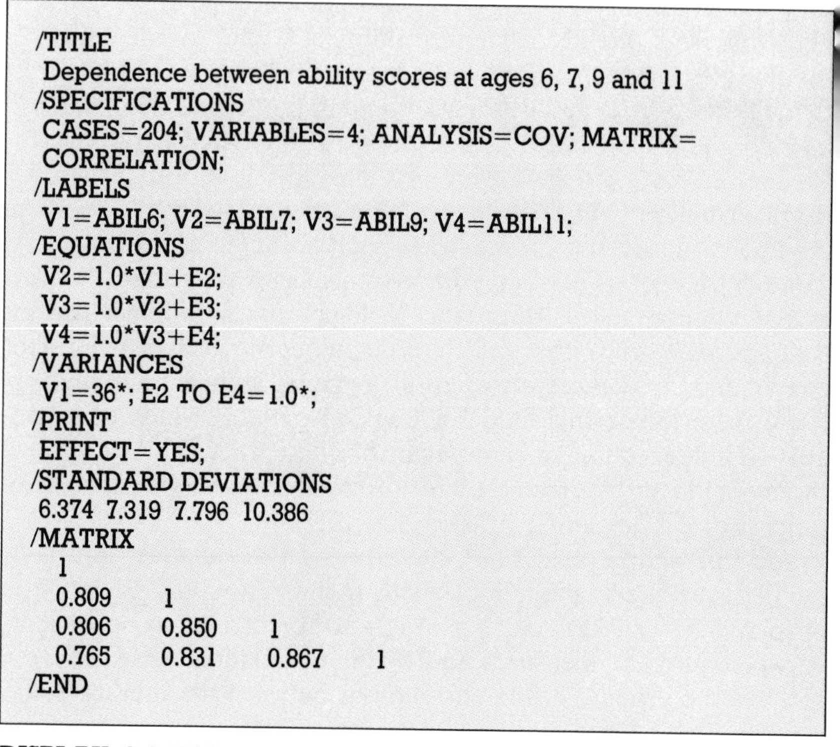

```
/TITLE
 Dependence between ability scores at ages 6, 7, 9 and 11
/SPECIFICATIONS
 CASES=204; VARIABLES=4; ANALYSIS=COV; MATRIX=
 CORRELATION;
/LABELS
 V1=ABIL6; V2=ABIL7; V3=ABIL9; V4=ABIL11;
/EQUATIONS
 V2=1.0*V1+E2;
 V3=1.0*V2+E3;
 V4=1.0*V3+E4;
/VARIANCES
 V1=36*; E2 TO E4=1.0*;
/PRINT
 EFFECT=YES;
/STANDARD DEVIATIONS
 6.374 7.319 7.796 10.386
/MATRIX
   1
   0.809    1
   0.806    0.850    1
   0.765    0.831    0.867    1
/END
```

DISPLAY 6.4 EQS command file for simple autoregressive model of continuity.

and 7, and 7 and 9. Since each r must be less than unity, the structure of this model implies that the expected correlation of a measure at time t with another one made s periods later (or earlier) will progressively decline as s increases. As we noted, this did appear to be a feature of the observed correlations in Display 6.1.

```
DECOMPOSITION OF EFFECTS WITH NONSTANDARDIZED VALUES

PARAMETER TOTAL EFFECTS
 ABIL7 = V2 =  .929*V1 + 1.000 E2
 ABIL9 = V3 =  .905*V2 +  .841 V1 + .905 E2 + 1.000 E3
ABIL11 = V4 = 1.046 V2 + 1.155*V3 + .971 V1 + 1.046 E2 + 1.155 E3 + 1.000 E4

DECOMPOSITION OF EFFECTS WITH NONSTANDARDIZED VALUES

PARAMETER INDIRECT EFFECTS
 ABIL9 = V3 =    .841 V1 +    .905 E2
                 .072         .063
               11.633       14.451
ABIL11 = V4 =  1.046 V2 +    .971 V1 +   1.046 E2 +   1.155 E3
                 .062         .063         .042         .047
               16.857       15.379       24.790       24.790

DECOMPOSITION OF EFFECTS WITH STANDARDIZED VALUES

PARAMETER TOTAL EFFECTS
 ABIL7 = V2 = .809*V1 + .588 E2
 ABIL9 = V3 = .850*V2 + .688 V1 + .500 E2 + .527 E3
ABIL11 = V4 = .737 V2 + .867*V3 + .596 V1 + .433 E2 + .457 E3 + .498 E4

DECOMPOSITION OF EFFECTS WITH STANDARDIZED VALUES

PARAMETER INDIRECT EFFECTS
 ABIL9 = V3 = .688 V1 + .500 E2
ABIL11 = V4 = .737 V2 + .596 V1 + .433 E2 + .457 E3

MAXIMUM LIKELIHOOD SOLUTION (NORMAL DISTRIBUTION THEORY)

STANDARDIZED SOLUTION:
 ABIL7 = V2 = .809*V1 + .588 E2
 ABIL9 = V3 = .850*V2 + .527 E3
ABIL11 = V4 = .867*V3 + .498 E4
```

DISPLAY 6.5 Decomposition of direct and indirect effects in the continuity of ability.

Display 6.4 gives the EQS instructions for fitting this model. The /PRINT subcommand EFFECT=YES, produces the additional output shown in Display 6.5. That output gives the estimates of the **total** and **indirect effects** of variables upon each other. The **direct effects** are the same as those given in the standard output and their interpretation has already been discussed (see, for example, Section 5.3). The indirect effects are new.

As we have seen, the indirect effects along a (valid) path be-

V1 to V3 = (V1 to V2)(V2 to V3) = 0.809 × 0.850 = 0.688

V2 to V4 = (V2 to V3)(V3 to V4) = 0.850 × 0.867 = 0.737

V1 to V4 = (V1 to V2)(V2 to V3)(V3 to V4)
= 0.809 × 0.850 × 0.867
= 0.596

DISPLAY 6.6 Calculation of indirect effects from results in Display 6.5.

tween two variables are obtained by multiplying all the direct effects along the path and the variances of any intervening variables. This is most straightforwardly seen in the case of the standardized solution, since, for such a solution, all the intervening variables have a variance of 1 (and the coefficients along the paths are correlation coefficients). Display 6.6 presents numerical examples that illustrate this procedure. For variables connected by more than one path, the indirect effects would have been obtained by summing the effects for each path.

Total effects are obtained by summing indirect and direct effects. Note that even for the standardized solution, the total effects of one variable on others, or of all other variables on one of them, can commonly exceed 1.

In fact, this model does not fit the data very well. It gives a 3 d.f. goodness-of-fit chi-squared statistic of 61.82 and a CFI of 0.928. The standardized residuals for the covariance of behaviour at 6 and 9, 6 and 11 and 7 and 9 are 0.118, 0.169 and 0.094, respectively, that is, the measures are more highly correlated than our model predicts. Some exploration of other models would seem to be required.

An obvious modification would be to allow a direct effect between age 6 ability and age 11 ability, since this is the relationship with the largest residual. This would rarely be the modification that a 'modeller' would pursue, unless there were some additional specific theoretical reason why this link should exist but not those between 6 and 9 and 7 and 9 as well. Rather, it is standard practice to exhaust models where effects are close in time before moving on to ones with more distant links. Thus one modification to the model would be to allow for so-called **second-order autoregressive** relationships, in other words direct effects not just from one meas-

```
/EQUATIONS
V2=1.0*V1+E2;
V3=1.0*V2+0.0*V1+E3;
V4=1.0*V3+0.0*V2+E4;
```

ISPLAY 6.7 EQS equations paragraph for a second-order autoregressive odel.

re to the next but further onward to the next but one. This would dd links between the measures at 6 and 9 and 7 and 11, as shown n Display 6.7.

This model fits much better, with a 1 d.f. chi-squared statistic of .87 and a CFI of 0.999. Moreover, the addition of the two extra irect effects between ages 6 and 9, and ages 7 and 11, has not only educed (to zero) the residuals for those specific elements of the orrelation matrix but has also accounted for a substantial proortion of the apparent longer-term link between ages 6 and 11, educing the standardized residual for this element from 0.169 to .024. How has this come about?

Display 6.8 gives the effect decomposition for this model. Look-

```
DECOMPOSITION OF EFFECTS WITH STANDARDIZED VALUES

PARAMETER TOTAL EFFECTS
 ABIL7 =V2 = .809*V1 + .588 E2
 ABIL9 =V3 = .573*V2 + .806*V1 + .337 E2 + .487 E3
ABIL11 =V4 = .671*V2 + .579*V3 + .741 V1 + .394 E2 + .282 E3 + .465 E4

DECOMPOSITION OF EFFECTS WITH STANDARDIZED VALUES

PARAMETER INDIRECT EFFECTS
 ABIL9 =V3 = .463*V1 + .337 E2
ABIL11 =V4 = .332*V2 + .741 V1 + .394 E2 + .282 E3

MAXIMUM LIKELIHOOD SOLUTION (NORMAL DISTRIBUTION
THEORY)

STANDARDIZED SOLUTION:
 ABIL7 =V2 = .809*V1 + .588 E2
 ABIL9 =V3 = .573*V2 + .343*V1 + .487 E3
ABIL11 =V4 = .339*V2 + .579*V3 + .465 E4
```

DISPLAY 6.8 Direct and indirect effects with second-order terms in the model.

ing at the effects for the standardized values and comparing them with those in Display 6.5, the total effect of V1 on V4 has increased from 0.596 to 0.741. This is because there are now three paths between V1 and V4, namely the original path V1–V2–V3–V4 and the two new paths V1–V3–V4 and V1–V2–V4. The sum of effects over these three paths

$$(0.809 \times 0.573 \times 0.579) + (0.343 \times 0.579) + (0.809 \times 0.339) = 0.741$$

gives the new estimate for the total effect.

Exercise 6.1

Modify the model that allows autoregressive effects between observed variables (Display 6.4) to fit the model that allows first-order effects and the third-order effect between age 6 and age 1[illegible] ability.

6.4 DIRECT AND INDIRECT EFFECTS BETWEEN FACTORS

On the evidence of fitting the model of Display 6.7 alone, it would be most unwise to conclude that ability can appear, go dormant and then reappear. There are other reasons why direct effects between chronologically adjacent observed variables may not account for the correlation between the more distant variables. One of these is measurement error. Several previous chapters have illustrated how measurement error can be allowed for by using multiple indicators to obtain factors. It is to this that we now turn our attention.

Perhaps the simplest approach is to argue that ability is a fixed trait and that variation over time arises solely from measurement error, suggesting the pure **measurement model** or fixed-trait model shown in Display 6.9. However, if the paths between variables are traced out, then the expected pattern of correlation between measures is quickly seen to be unlike our observed matrix. Going from V1 gives expected correlations of V1 with V2 of r_1r_2, of V1 with V3 of r_1r_3 and of V1 with V4 of r_1r_4, where the rs in this case are measures of the reliability of the ability measure at each age. Thus unless the reliability declines with age, measured ability at age 6

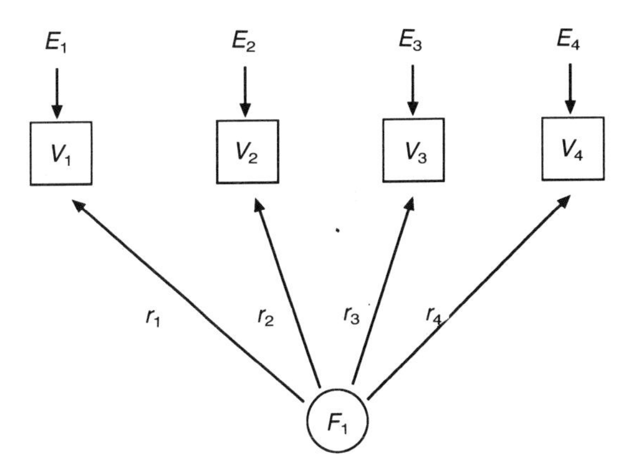

DISPLAY 6.9 Fixed-trait model of ability measured subject to error.

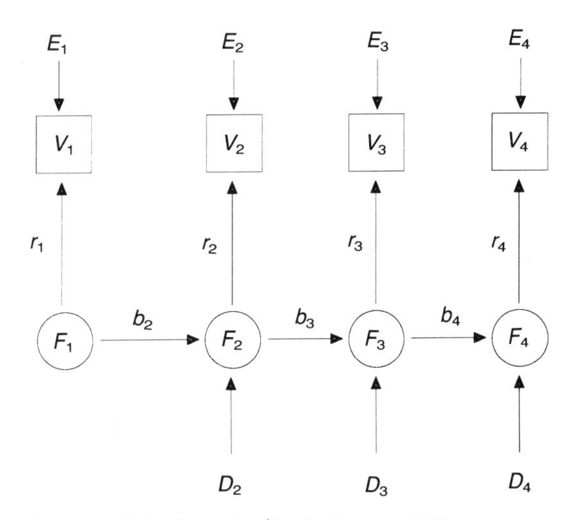

DISPLAY 6.10 A model of continuity in latent ability.

would be expected to correlate as highly with that at 11 as it does with that at 7, a pattern we do not observe. The reader is left to confirm more formally the inappropriateness of this one-factor model.

A more plausible model is shown in Display 6.10. This shows a similar model to that in Display 6.2, but now the autoregressive structure occurs between the latent factors rather than directly between the observed variables. This is the first example of a **structural model** involving postulated causal effects between one

```
/TITLE
 Dependence between ability scores at ages 6, 7, 9 and 11
 Simple latent variable model
/SPECIFICATIONS
 CASES=204; VARIABLES=4; ANALYSIS=COV;
 MATRIX=
 CORRELATION;
/LABELS
 V1=ABIL6; V2=ABIL7; V3=ABIL9; V4=ABIL11;
/EQUATIONS
 V1=F1+E1;
 V2=F2+E2;
 V3=F3+E3;
 V4=F4+E4;
 F2=1.0*F1+D2;
 F3=1.0*F2+D3;
 F4=1.0*F3+D4;
/VARIANCES
 F1=30*; D2 TO D4=1.0*; E1 TO E4=1.0*;
/CONSTRAINTS
 (E1,E1)=(E2,E2)=(E3,E3)=(E4,E4);
/PRINT
 EFFECT=YES;
/STANDARD DEVIATIONS
 6.374 7.319 7.796 10.386
/MATRIX
 1
 0.809    1
 0.806    0.850    1
 0.765    0.831    0.867    1
/END
```

DISPLAY 6.11 Simple model for the continuity of latent ability.

latent variable and another. Display 6.11 shows the EQS commands for this model. The first four equations are familiar measurement equations, and the model constrains the errors in measuring ability at ages 6, 7, 9 and 11 to have the same variance and to be uncorrelated. The variance in latent ability at age 6 is directly estimated, while those at ages 7, 9 and 11 are predicted by three equations involving the previous latent ability and independent disturbance terms D2 to D4. The presence of these disturbance terms means that latent ability is not an entirely fixed and stable trait, but may vary, increasing or decreasing relative to other children, from one time to the next.

```
DECOMPOSITION OF EFFECTS WITH NONSTANDARDIZED VALUES

PARAMETER TOTAL EFFECTS
ABIL6  =V1 =   1.000 F1 +    1.000 E1
ABIL7  =V2 =   1.000 F2 +   -.000 F3 +  1.120 F1 +  1.000 E2 + 1.000 D2 + -.000 D3
ABIL9  =V3 =   1.044 F2 +   1.000 F3 +  1.170 F1 +  1.000 E3 + 1.044 D2 +  1.000 D3
ABIL11 =V4 =   1.353 F2 +   1.296 F3 +  1.000 F4 +  1.516 F1 + 1.000 E4 +  1.353 D2 + 1.296 D3 + 1.000 D4
F2     =F2 =   -.000 F3 +   1.120*F1 +  1.000 D2 + -.000 D3
F3     =F3 =   1.044*F2 +   -.000 F3 +  1.170 F1 +  1.044 D2 + 1.000 D3
F4     =F4 =   1.353 F2 +   1.296*F3 +  1.516 F1 +  1.353 D2 + 1.296 D3 +  1.000 D4

DECOMPOSITION OF EFFECTS WITH NONSTANDARDIZED VALUES

PARAMETER INDIRECT EFFECTS
ABIL7  =V2 =   -.000 F3 +   1.120 F1 +  1.000 D2 + -.000 D3
                .000         .064
              -:0.000      17.391
ABIL9  =V3 =   1.044 F2 +   -.000 F3 +  1.170 F1 +  1.044 D2 + 1.000 D3
                .046         .000        .067
              22.590       -:0.000     17.391
ABIL11 =V4 =   1.353 F2 +   1.296 F3 +  1.516 F1 +  1.353 D2 + 1.296 D3 +  1.000 D4
                .068         .054        .087
              20.036       23.900      17.391
F2     =F2 =   -.000 F3 +   -.000 D3
                .000         .073
              -:0.000       -.000
F3     =F3 =   -.000 F3 +   1.170 F1 +  1.044 D2 + -.000 D3
                .000         .101        .073        .070
              -:0.000      11.595      14.362       -.000
F4     =F4 =   1.353 F2 +   1.516 F1 +  1.353 D2 + 1.296 D3
                .068         .111        .057        .054
              20.036       13.608      23.900      23.900

DECOMPOSITION OF EFFECTS WITH STANDARDIZED VALUES

PARAMETER TOTAL EFFECTS
ABIL6  =V1 =    .912 F1 +    .410 E1
ABIL7  =V2 =    .934 F2 +   -.000 F3 +   .891 F1 +   .358 E2 +  .280 D2 + -.000 D3
ABIL9  =V3 =    .914 F2 +    .942 F3 +   .872 F1 +   .335 E3 +  .274 D2 +   .231 D3
ABIL11 =V4 =    .889 F2 +    .917 F3 +   .968 F4 +   .849 F1 +  .252 E4 +   .267 D2 +  .225 D3 +  .308 D4
F2     =F2 =   -.000 F3 +    .954*F1 +   .300 D2 +  -.000 D3
F3     =F3 =    .970*F2 +   -.000 F3 +   .925 F1 +   .291 D2 +  .245 D3
F4     =F4 =    .919 F2 +    .948*F3 +   .877 F1 +   .275 D2 +  .232 D3 +   .318 D4

DECOMPOSITION OF EFFECTS WITH STANDARDIZED VALUES

PARAMETER INDIRECT EFFECTS
ABIL7  =V2 =   -.000 F3 +    .891 F1 +   .280 D2 +  -.000 D3
ABIL9  =V3 =    .914 F2 +   -.000 F3 +   .872 F1 +   .274 D2 +  .231 D3
ABIL11 =V4 =    .889 F2 +    .917 F3 +   .849 F1 +   .267 D2 +  .225 D3 +   .308 D4
F2     =F2 =   -.000 F3 +   -.000 D3
F3     =F3 =   -.000 F3 +    .925 F1 +   .291 D2 +  -.000 D3
F4     =F4 =    .919 F2 +    .877 F1 +   .275 D2 +   .232 D3

MAXIMUM LIKELIHOOD SOLUTION (NORMAL DISTRIBUTION THEORY)

STANDARDIZED SOLUTION:
ABIL6  =V1 =    .912 F1 +    .410 E1
ABIL7  =V2 =    .934 F2 +    .358 E2
ABIL9  =V3 =    .942 F3 +    .335 E3
ABIL11 =V4 =    .968 F4 +    .252 E4
F2     =F2 =    .954*F1 +    .300 D2
F3     =F3 =    .970*F2 +    .245 D3
F4     =F4 =    .948*F3 +    .318 D4
```

DISPLAY 6.12 Direct and indirect effects in the continuity of latent ability.

Unlike the model with just first-order autoregressive effects between the *observed* ability scores this model fits well, giving a 2 d.f. chi-squared statistic of 1.43, a CFI of 1.00 and no standardized residual larger than 0.01. Display 6.12 gives the effect decomposition for this model, together with the standardized solution. What is clear is that there is substantially greater continuity in ability, as shown by the effects between the factors, than the simple correlations in the observed measures suggest.

Why should the presence of measurement error have given rise to the impression that 'sleeper effects' were present, as represented in a more complicated structural model involving second-order autoregressive effects? Some intuitive understanding is possible if one considers the impact of measurement error on prediction. Consider two individuals, one scoring high at both 6 and 7 and the other high at 7 but low at 6. For the prediction of ability at 9, using the observed score at 7 alone (as the first-order autoregressive model does), these two individuals will be assigned the same predicted score. However, when one considers measurement error, the individual with high scores at both 6 and 7 probably does have a high ability, while the one that scores inconsistently probably has a more intermediate ability. Including the age 6 observed score in the equation to predict age 9 score helps to distinguish these two individuals, but in so doing leaves the impression not that they might in fact have different age 7 abilities but that there are direct effects due to age 6 score. The latent variable model deals with the measurement errors directly and in so doing allows the simple structural model beneath to show through.

6.5 MORE ON THE EFFECTS OF MEASUREMENT ERROR AND IDENTIFIABILITY

The previous model was estimated under the constraint that measurement error variances remained constant with age. What if this constraint is removed? Display 6.13 shows the expected pattern of correlations that would then arise, the r coefficients being the correlations between latent scores and their corresponding observed score, and the b coefficients being the correlations of a latent ability score with the previous latent ability score. If the scores are measured with equal reliability, then the correlation between observed scores will decline progressively away from the diagonal.

	V1	V2	V3	V4
V1	1			
V2	$r_1b_2r_2$	1		
V3	$r_1b_2b_3r_3$	$r_2b_3r_3$	1	
V4	$r_1b_2b_3b_4r_4$	$r_2b_3b_4r_4$	$r_3b_4r_4$	1

DISPLAY 6.13 Expected correlation matrix from latent ability model in Display 6.10.

```
PARAMETER          CONDITION CODE
E4,E4              LINEARLY DEPENDENT ON OTHER PARAMETERS
F2,F1              LINEARLY DEPENDENT ON OTHER PARAMETERS

MAXIMUM LIKELIHOOD SOLUTION (NORMAL DISTRIBUTION THEORY)
E4,E4              VARIANCE OF PARAMETER ESTIMATE IS SET TO ZERO.
F2,F1              VARIANCE OF PARAMETER ESTIMATE IS SET TO ZERO.

MAXIMUM LIKELIHOOD SOLUTION (NORMAL DISTRIBUTION THEORY)

*** WARNING *** TEST RESULTS MAY NOT BE APPROPRIATE DUE TO
CONDITION CODE

GOODNESS OF FIT SUMMARY

INDEPENDENCE MODEL CHI-SQUARE = 820.575, BASED ON 6 DEGREES OF
FREEDOM

INDEPENDENCE AIC = 808.57548 INDEPENDENCE CAIC = 782.66676
       MODEL AIC =   3.14022        MODEL CAIC =   7.45834

CHI-SQUARE = 1.140 BASED ON -1 DEGREES OF FREEDOM
NONPOSITIVE DEGREES OF FREEDOM. PROBABILITY COMPUTATIONS ARE
UNDEFINED.

BENTLER-BONETT NORMED FIT INDEX = .999

NON-NORMED FIT INDEX WILL NOT BE COMPUTED BECAUSE A DEGREES
OF FREEDOM IS ZERO.

MAXIMUM LIKELIHOOD SOLUTION (NORMAL DISTRIBUTION THEORY)

STANDARDIZED SOLUTION:
ABIL6  =V1 =  .988 F1 +  .157 E1
ABIL7  =V2 =  .929 F2 +  .370 E2
ABIL9  =V3 =  .946 F3 +  .323 E3
ABIL11 =V4 = 1.000 F4 + 1.000 E4
F2     =F2 =  .882*F1 +  .471 D2
F3     =F3 =  .971*F2 +  .241 D3
F4     =F4 =  .916*F3 +  .401 D4
```

DISPLAY 6.14 Selected EQS output for under-identified latent ability model.

However, if one variable, say at age 9, happened to be very poorly measured then the correlation between age 7 and 11 observed scores could be larger than those between ages 7 and 9 and ages 9 and 11. Thus measurement error does not simply attenuate the effects that one might estimate but can, and often does, result in needlessly complex or incorrect relationships being found.

Can this unrestricted model be estimated? Selected parts of the output given by EQS for this model are shown in Display 6.14. Note the goodness-of-fit chi-squared statistic with negative degrees of freedom and the two warning messages (that the variance of E4 was constrained at lower bound and that the effect of F1 on F2 was linearly dependent upon other parameters). These are both due to the model not being *identified*, that is, there are more parameters being estimated than there are observed variances and covariances.

Some simple algebraic manipulation in the lower half of Display 6.13 shows that some parameters, however, can be *uniquely* identified. From the expected correlation matrix we have

$$\mathrm{Cor}(V2,V1)\mathrm{Cor}(V3,V2)/\mathrm{Cor}(V3,V1) = r_1 b_2 r_2 . r_2 b_3 r_3 / r_1 b_2 b_3 r_3 = r_2^2$$

and

$$\mathrm{Cor}(V4,V3)\mathrm{Cor}(V3,V2)/\mathrm{Cor}(V4,V2) = r_3 b_4 r_4 . r_2 b_3 r_3 / r_2 b_3 b_4 r_4 = r_3^2$$

Given estimates of r_2 and r_3, an estimate of b_3 can be obtained from Cor(V3,V2) $= r_2 b_3 r_3$. Thus the middle part of the model in Display 6.10 can be identified without imposing measurement restrictions and without having more than one indicator for each latent variable. However, none of the other parameters is identified. This illustrates that questions of identification are often not straightforward.

The problem can be explored empirically by imposing constraints on the left- and right-hand ends of this model and examining the impact on the solution obtained. We could constrain the error variances at time 1 (E1,E1) and time 4 (E4,E4) to be zero (in fact EQS had constrained this last variance to zero anyway in order to come up with its previous 'answer'). Fitting this constrained model gives a goodness-of-fit chi-squared statistic of 1.14 with 1 d.f. (CFI equal to 1.00) and the standardized solution shown in Display 6.15. The effect of F2 on F3 remains unchanged from the previous model at 0.971, since it was identified anyway. The estimate of the

```
MAXIMUM LIKELIHOOD SOLUTION (NORMAL DISTRIBUTION THEORY)
STANDARDIZED SOLUTION:
ABIL6   =V1 = 1.000 F1
ABIL7   =V2 =  .929 F2  + .370 E2
ABIL9   =V3 =  .946 F3  + .323 E3
ABIL11  =V4 = 1.000 F4
F2      =F2 =  .871*F1  + .491 D2
F3      =F3 =  .971*F2  + .241 D3
F4      =F4 =  .916*F3  + .401 D4
```

DISPLAY 6.15 Selected EQS output for the just-identified latent ability model.

effect of F1 on F2 has changed, while the estimate of F3 on F4 is unaltered only because the constraint that the variance of E4 was 0 had already been imposed by EQS.

6.6 IMPOSING CONSTRAINTS: MODELLING CONTINUITY IN EQUALLY SPACED OBSERVATIONS

It is often interesting to test hypotheses about the pattern of development of some measure over time. Is the development of cognitive ability, for example, uniform or are there particular periods of discontinuity or greater disturbance in the relative abilities of individuals, perhaps associated with changing schools or puberty?

Where the measures are equally spaced in time, it is straightforward to add constraints that impose some degree of uniform progression. For the model of Display 6.10 (assuming, for the moment, all intervals to be equal), the coefficients denoted b_2 to b_4 could all be constrained equal, as could the variances of the period-specific disturbances d_2 to d_4. The necessary additions to the EQS command file of Display 6.11 would leave the constraints paragraph as shown in Display 6.16. Fitting this model gives a 6

```
/CONSTRAINTS
 (E1,E1)=(E2,E2)=(E3,E3)=(E4,E4);
 (F2,F1)=(F3,F2)=(F4,F3);
 (D2,D2)=(D3,D3)=(D4,D4);
```

DISPLAY 6.16 Constraints to impose uniform continuity over time.

d.f. goodness-of-fit chi-squared statistic of 24.14 (CFI equal to 0.978).

6.7 IMPOSING CONSTRAINTS: MODELLING CONTINUITY IN UNEQUALLY SPACED OBSERVATIONS

Of course, these data are not equally spaced in time. The interval between the first and second measures is one year, while those between the second and third measures and between the third and fourth measures are two years. What should the appropriate constraints look like? This is most easily determined by redrawing the path diagram, with the 'missing' measures at ages 8 and 10 inserted as in Display 6.17. Tracing the paths shows that the effective coefficient between measures two years apart is obtained by multiplying the two coefficients for each one-year interval. The hypothesis of uniform coefficients over time thus does not imply simple linear constraints over unequally spaced intervals (halving the coefficient for doubling the time interval) but non-linear con-

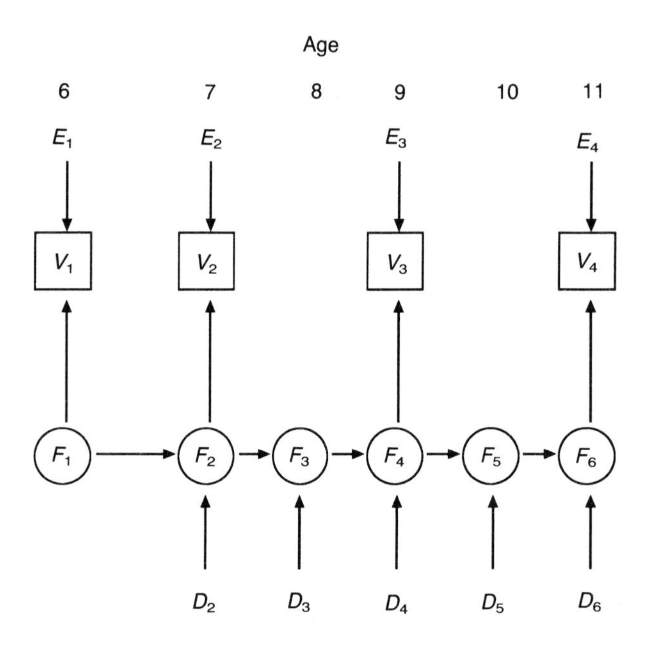

DISPLAY 6.17 A model of uniform continuity in latent ability for unequally spaced intervals.

straints, raising the coefficient to the power of the length of the interval (squaring the coefficient for doubling the time interval).

Such non-linear constraints cannot be imposed directly within EQS, but the model of Display 6.17 can, in fact, be fitted by introducing the 'missing' factors directly into the model. This is

```
/TITLE
 Dependence between ability scores at ages 6, 7, 9 and 11 – Uniform
 Development Model
/SPECIFICATIONS
 CASES=204; VARIABLES=4; ANALYSIS=COV; MATRIX=
 CORRELATION;
/LABELS
 V1=ABIL6; V2=ABIL7; V3=ABIL9; V4=ABIL11;
/EQUATIONS
 V1=F1+E1;
 V2=F2+E2;
 V3=F4+E3;
 V4=F6+E4;
 F2=1.0*F1+D2;
 F3=1.0*F2+D3;
 F4=1.0*F3+D4;
 F5=1.0*F4+D5;
 F6=1.0*F5+D6;
/VARIANCES
 F1=30*; D2 TO D6=1.0*; E1 TO E4=1.0*;
/CONSTRAINTS
 (E1,E1)=(E2,E2)=(E3,E3)=(E4,E4);
 (F2,F1)=(F3,F2)=(F4,F3)=(F5,F4)=(F6,F5);
 (D2,D2)=(D3,D3)=(D4,D4)=(D5,D5)=(D6,D6);
/PRINT
 EFFECT=YES;
/STANDARD DEVIATIONS
6.374 7.319 7.796 10.386
/MATRIX
 1
 0.809   1
 0.806   0.850   1
 0.765   0.831   0.867   1
/END
```

DISPLAY 6.18 Uniform change in latent ability: a model for unequal intervals.

```
MAXIMUM LIKELIHOOD SOLUTION (NORMAL DISTRIBUTION
THEORY)

CONSTRUCT EQUATIONS WITH STANDARD ERRORS AND TEST
STATISTICS

F2 =F2 =   1.090*F1 + 1.000 D2
            .012
          88.231
F3 =F3 =   1.090*F2 + 1.000 D3
            .012
          88.231
F4 =F4 =   1.090*F3 + 1.000 D4
            .012
          88.231
F5 =F5 =   1.090*F4 + 1.000 D5
            .012
          88.231
F6 =F6 =   1.090*F5 + 1.000 D6
            .012
          88.231

MAXIMUM LIKELIHOOD SOLUTION (NORMAL DISTRIBUTION
THEORY)

VARIANCES OF INDEPENDENT VARIABLES

V          F                       E                        D
   I F1 - F1     33.107*I E1 -ABIL6       7.834*I D2 - F2      2.399*
   I              3.764 I                  .872 I               .738
   I              8.795 I                 8.983 I              3.253
   I                    I                       I
   I                    I E2 -ABIL7       7.834*I D3 - F3      2.399*
   I                    I                  .872 I               .738
   I                    I                 8.983 I              3.253
   I                    I                       I
   I                    I E3 -ABIL9       7.834*I D4 - F4      2.399*
   I                    I                  .872 I               .738
   I                    I                 8.983 I              3.253
   I                    I                       I
   I                    I E4 -ABIL11      7.834*I D5 - F5      2.399*
   I                    I                  .872 I               .738
   I                    I                 8.983 I              3.253
   I                    I                       I
   I                    I                       I D6 - F6      2.399*
   I                    I                       I               .738
   I                    I                       I              3.253
   I                    I                       I
```

DISPLAY 6.19 Selected EQS output from the model of uniform change in latent ability.

```
MAXIMUM LIKELIHOOD SOLUTION (NORMAL DISTRIBUTION
THEORY)

STANDARDIZED SOLUTION:
ABIL6  =V1 = .899 F1 + .437 E1
ABIL7  =V2 = .918 F2 + .397 E2
ABIL9  =V3 = .944 F4 + .330 E3
ABIL11 =V4 = .962 F6 + .275 E4
F2     =F2 = .971*F1 + .240 D2
F3     =F3 = .977*F2 + .215 D3
F4     =F4 = .981*F3 + .193 D4
F5     =F5 = .985*F4 + .174 D5
F6     =F6 = .987*F5 + .158 D6
```

DISPLAY 6.19 *Continued*

shown in Display 6.18, in which factors F3 and F5 are 'phantoms' having no corresponding observed measure. However, because of the constraints on the disturbance terms and continuity coefficients, no new parameters have to be estimated and the model remains identified. Thus, this model, which correctly imposes uniform development over unequal intervals, also has 6 degrees of freedom, and in fact appears to fit rather better than the previous one, giving a smaller χ^2 of 19.38 and a CFI of 0.984.

Display 6.19 gives the construct equations, independent variable variances and standardized solution for this model. As can be seen from the construct equations and variances, the constraints have been correctly imposed. However, the standardized solution shows that although the correlations between the factors are very similar they are not identical. This is because the variances of the dependent factors F2 to F6 have not been constrained equal in the estimated model.

Exercise 6.2

Fit and then examine the residuals from the autoregressive factor model for the uniform development of ability for observations unequally spaced in time (Display 6.18). How might the model be relaxed to achieve a better fit?

6.8 EXPLAINING GROWTH IN VARIANCE: AN ALTERNATIVE MODEL

The autoregressive structure of the models examined so far is by far the most common form of structural equation model applied to longitudinal data. Within this model the main focus of attention appears to be the explanation of the pattern of correlation in measures over time. However, returning to the summary statistics of Display 6.1, a quite different but striking feature of the data is the steady growth in the variance of the ability measure. The scores must be fanning out as age increases, the difference between the better and worse children increasing with time.

One way of thinking about the process of gaining ability is one in which, during each interval between measures, each child adds an amount of new ability, and these increments (or possibly decrements) are 'frozen in', accumulating over time. Such a model is shown in Display 6.20. Factor F1 essentially defines each child's initial score at age 6. This intial factor contributes directly to all subsequent scores. During each subsequent interval, increments in scores are obtained corresponding to the factors F2, F3 and F4, each also contributing to subsequent scores. To identify the model it will usually be necessary to assume that the increments between intervals are independent one from another. In other words, a

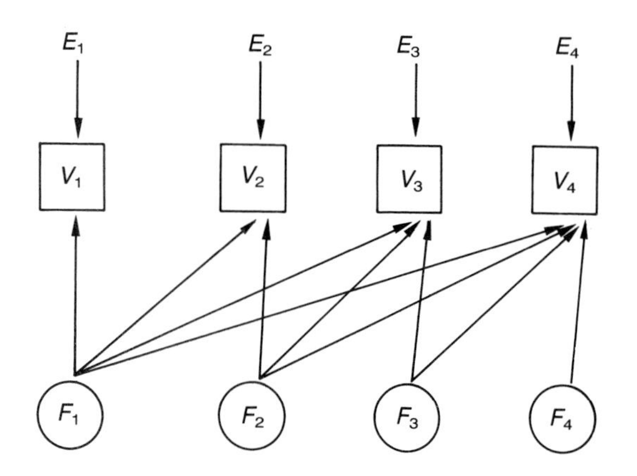

DISPLAY 6.20 A model of ability change as a random walk.

```
/TITLE
 Dependence between ability scores at 6, 7, 9 and 11
 Random walk model of latent ability
/SPECIFICATIONS
 CASES=204; VAR=4; ANALYSIS=COVARIANCE; MATRIX=
 CORRELATIONS;
/LABELS
 V1=ABIL6; V2=ABIL7, V3=ABIL9; V4=ABIL11;
/EQUATIONS
 V1=F1+E1;
 V2=F1+F2+E2;
 V3=F1+F2+F3+E3;
 V4=F1+F2+F3+F4+E4;
/VARIANCES
 F1 TO F4=25.0*;
 E1 TO E4=8*;
/CONSTRAINTS
 (E1,E1)=(E2,E2)=(E3,E3)=(E4,E4);
/MATRIX
 1
 0.809    1
 0.806    0.850    1
 0.765    0.831    0.867    1
/STANDARD DEVIATIONS
 6.374    7.319    7.796    10.386
/END
```

DISPLAY 6.21 Random walk model for change in latent ability.

large increase in one interval does not imply an increased chance of a large increment in the next. This gives rise to the description of this model as a **random walk** or **Wiener** model.

Display 6.21 gives the EQS command file for this model. Perhaps to some initial surprise, this model gives a CFI of 0.956. A number of authors would deem this CFI value as acceptable, although the 5 d.f. goodness-of-fit chi-squared statistic of 41.21 is less flattering. Even so, to fit this well the model must also be explaining both the growth in variance and the pattern of correlations. It does the latter through the fact that all previous factors are components of variance of all later ability factors.

Exercise 6.3

The random walk or Wiener model assumes that the increments to ability during each interval are independent. What sort of correlation structure might be more plausible? Try fitting such a model.

6.9 DETERMINING CAUSAL DIRECTION

In the previous examples the question of causal direction was not at issue – effects backward in time were theoretically implausible. However, with more than one variable measured at each time point, there are many situations where the question as to which influences the other is an open one. Such is the case for the data shown in Display 6.22 on the reading scores and teachers' ratings of behaviour for 362 boys living in inner London, measured at the ages of 10, 13 and 14. Problems in reading could make school work unrewarding and give rise to difficult behaviour in class. Alternatively, behaviour problems could give rise to poor reading through the child's failing to pay attention in class (or failing to attend class at all).

The behaviour score is the total score on n items, each scored between 0 and 2. A large number of children score 0 on all items. Taking the square root of these scores (before computing means, standard deviations and correlations) makes the data approximate multivariate normality a little more closely. The reading scores are essentially standardized scores that would be expected to have a mean of around 100 and standard deviation of around 15. These scores already approximate normality well.

Display 6.23 illustrates the various effects that could be postu-

	Beh10	Read10	Beh13	Read13	Beh14	Read14
Beh10	1					
Read10	−0.3054	1				
Beh13	0.3868	−0.2801	1			
Read13	−0.2722	0.7485	−0.3166	1		
Beh14	0.2094	−0.1479	0.4819	−0.2031	1	
Read14	−0.3012	0.7434	−0.3705	0.8644	−0.1950	1
Stan.Dev.:	1.3856	13.3764	1.3708	13.1558	1.3542	13.5728

DISPLAY 6.22 Reading and behaviour at ages 10, 13 and 14: correlation matrix.

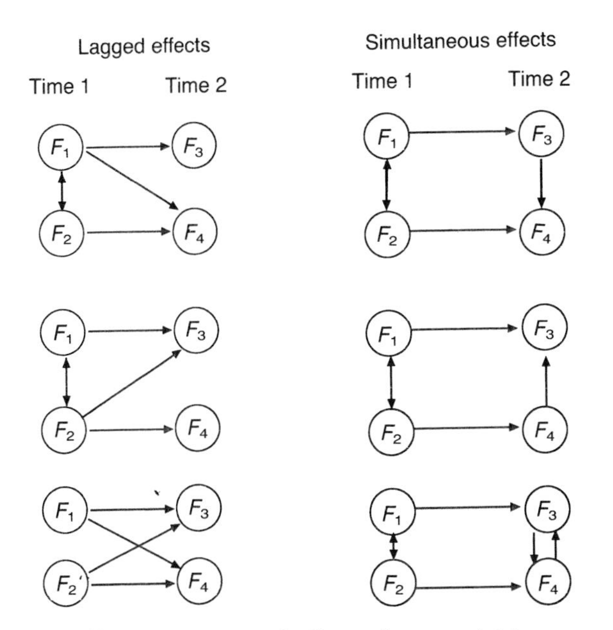

DISPLAY 6.23 Different patterns of effect of two variables on each other.

lated between the two variables measured at two time points. The basic structure is to allow the two variables to be freely correlated at time 1 and then to attempt to predict their variance and covariance at time 2 through allowing only a subset of the diagonal and vertical effects shown.

In fact, with only four observed variables there are, in general, too few observed variances and covariances to fit the desired models. We have six variables (measures of behaviour and reading at 10, 13 and 14), and have chosen to use those at ages 13 and 14 as indicators of latent scores at time 2, as shown in Display 6.24.

The EQS command file of Display 6.25 fits the model with simultaneous effects at time 2 from reading to behaviour and behaviour to reading. Measurement error variances have been assumed constant over time and factor loadings equal for both time 2 measures. The model gave a goodness-of-fit chi-squared statistic of 26.35 with 10 degrees of freedom and a CFI of 0.984. Display 6.26 gives the results from the construct equations and standardized results sections of the output. These suggest significant effects for poor behaviour on reading ($z = -2.181$) but non-significant effects for reading on behaviour ($z = -0.992$).

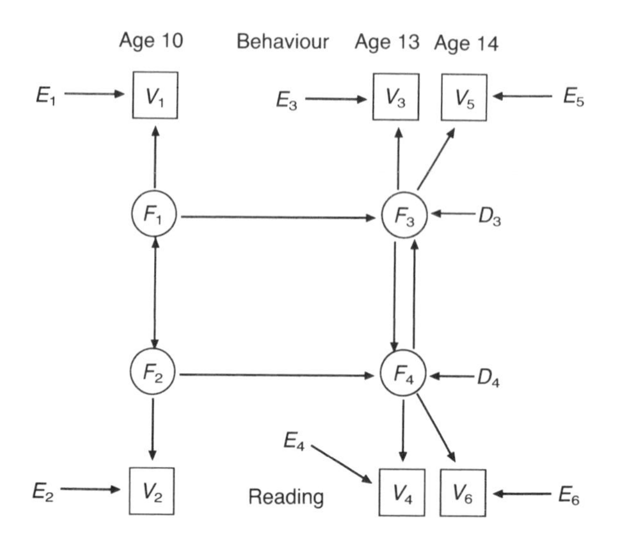

DISPLAY 6.24 Detailed model of effects between reading and behaviour measured on three occasions.

However, examination of the standardized solution indicates that the relative sizes of these two estimated effects are similar so that an unequivocal answer should not be given. The reader would be advised to rerun the model twice, in each case restricting one of the two effects between reading and behaviour to zero (by removing an asterisk from one of the last two equations). Subtraction of the goodness-of-fit chi-squared values of these models from the 26.35 of the current model would then give likelihood ratio chi-squared tests for the removal of each effect, a useful check on the adequacy of the z-tests.

For many readers the ability to estimate simultaneous effects between variables may seem rather remarkable. It should be emphasized that this is achieved only through making some fairly stiff assumptions. In particular, what has been specifically excluded is any covariance between the disturbance terms D3 and D4. This implies that it has been assumed that nothing between time 1 and time 2 influences both reading and behaviour – any covariation in these variables, other than that accounted for by their time 1 correlation and the dependence of each variable on its prior score, is put down to the direct effects of one variable on the other. Researchers have to evaluate the appropriateness of such assumptions for their particular problem.

```
/TITLE
 Analysis of Behavioural and Reading Data
/SPECIFICATIONS
 CASES=362; VARS=6; ANAL=COV; MATRIX=COR;
/LABELS
 V1=b10; V2=r10; V3=b13; V4=r13; V5=b14; V6=r14;
/EQUATIONS
 V1=F1+E1;
 V2=F2+E2;
 V3=F3+E3;
 V4=F4+E4;
 V5=F3+E5;
 V6=F4+E6;
 F3=0.3*F1-0.02*F4+D3;
 F4=0.74*F2-1.5*F3+D4;
/VARIANCES
 E1 TO E6=1.0*;
 D3=0.64*; D4=49.0*;
 F1=2.0*; F2=179.0*;
/COVARIANCES
 F1,F2=-5.6*;
/CONSTRAINTS
 (E1,E1)=(E3,E3)=(E5,E5);
 (E2,E2)=(E4,E4)=(E6,E6);
/MATRIX
  1.0000
 -0.3054   1.0000
  0.3868  -0.2801   1.0000
 -0.2722   0.7485  -0.3166   1.0000
  0.2094  -0.1479   0.4819  -0.2031   1.0000
 -0.3012   0.7434  -0.3705   0.8644  -0.1950   1.0000
/STANDARD DEVIATIONS
 1.3856 13.3764 1.3708 13.1558 1.3542 13.5728
/END
```

DISPLAY 6.25 A model with simultaneous effects between latent behaviour and latent reading ability.

Exercise 6.4

Within the analysis of the interdependence of reading and behaviour scores, fit the models that allow for cross-lagged effects between the factors rather than simultaneous effects.

```
F3  =F3 =  -.007*F4 +   .538*F1 + 1.000 D3
             .007         .120
            -.992        4.499
F4  = F4 = -1.584*F3 +   .821*F2 + 1.000 D4
             .726         .043
           -2.181       18.902

STANDARDIZED SOLUTION:
B10 =V1 =    .708 F1 +   .706 E1
R10 =V2 =    .930 F2 +   .368 E2
B13 =V3 =    .696 F3 +   .718 E3
R13 =V4 =    .930 F4 +   .368 E4
B14 =V5 =    .696 F3 +   .718 E5
R14 =V6 =    .930 F4 +   .368 E6
F3  = F3 =  -.098*F4 +   .558*F1 +  .785 D3
F4  = F4 =  -.121*F3 +   .822*F2 +  .488 D4
```

DISPLAY 6.26 Construct equations with standard errors and test statistics.

6.10 MODELLING MEANS AND COVARIANCES

In Chapters 1 and 2, manual calculation showed how, for a simple regression model, both the slope and the intercept can be estimated from sample means and covariances. However, the estimation of the slope did not require estimating the constant, and the answer obtained did not change when the intercept was estimated. This is quite a common situation with data from just a single group, that is, the additional complexity of modelling the means can leave unchanged the estimates of how the data covary. Thus modelling the means is more often encountered with multigroup analyses and is more fully illustrated in Chapter 7.

However, there are two situations when single-group estimates for the covariation in the data will change when the means are modelled as well. The first is where the model for the means is over-identified. This will typically result in forcing the estimate of the intercept of some regression to be rather different than that from a free regression. With the intercept pushed to some other value the estimated slope is likely to make compensating adjustments, illustrated in Display 6.27. An example of this situation is given in Exercise 6.6. Secondly, and perhaps most interestingly, some models attempt to explain the variation in means and covari-

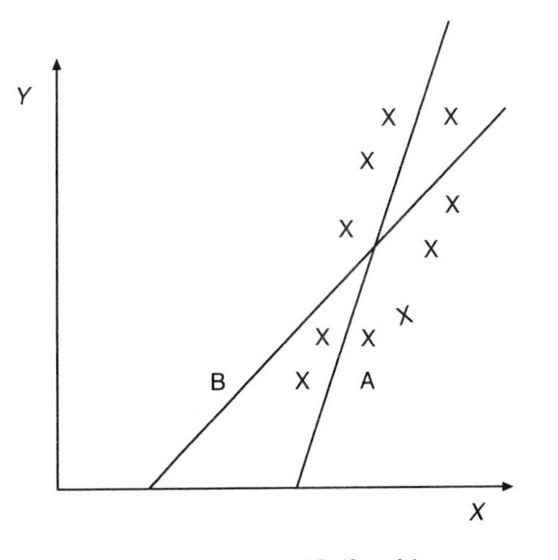

A=estimated line with 'free' intercept
B=estimated line with restricted intercept

DISPLAY 6.27 Effect of restricting intercept (or mean) on estimated slope (or covariance).

ances by a common model. In such models some parameters cannot be exclusively classified as involved in explaining means or covariances, but do both. The example that follows illustrates this.

6.11 STRUCTURED MEANS IN ONE GROUP: MODELLING LATENT GROWTH

The autoregressive models that began this chapter focused on explaining the correlation of children's ability scores over time. They were followed by a Wiener model that focused on the increasing variance of those scores. We now illustrate a **growth curve** model, which focuses simultaneously on correlations over time, on increases in variance and on shifts in mean values.

The path diagram is illustrated in Display 6.28. The model views the process of gaining ability as one determined by two latent characteristics of each child. The first latent characteristic is an initial ability at entry to the study, indicated by the factor F1. Fixing the path from F1 to V1 at 1 results in the estimate of the variance of F1 being the estimated variance of latent ability in the sample of children at age 6. It also results in the estimate of the

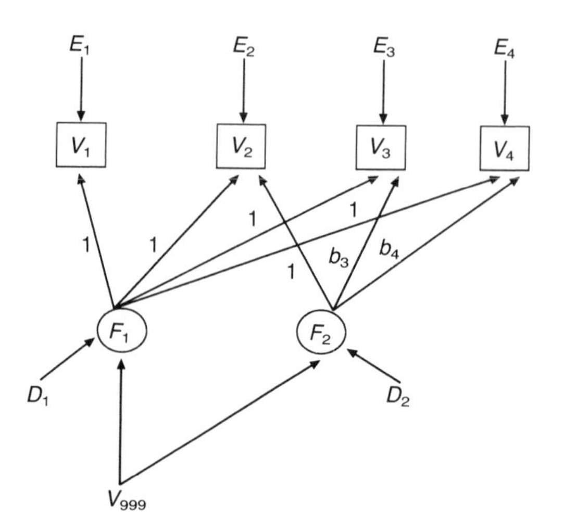

DISPLAY 6.28 A model of growth in latent ability.

coefficient from V999 to F1 being the estimate of the mean latent ability score at age 6. This initial ability is assumed 'frozen in', so that it remains as a component of all later scores (the coefficients from F1 to V2, F1 to V3 and F1 to V4 are all fixed to 1). The second latent characteristic is the slope of each child's trajectory over time, indicated by the factor F2. This is concerned with the relative progress of each child after age 6. Fixing the path from F2 to V2 to 1 results in the estimated coefficient from V999 to F2 being the mean increase in ability from age 6 to age 7. The estimate of coefficient b_3 for the path from F2 to V3 describes the mean progress from age 6 to age 9 as a multiple of the progress from age 6 to age 7, and similarly for the coefficient b_4.

The variance of the factor F2 is the variation between children in the slopes of their trajectories. This generates an expected pattern of scores as shown in Display 6.29, with the initial variance in ability determined by F1 and the extent of subsequent fanning-out by the variance of F2. If the two latent factors, F1 and F2, are independent, then the expected variance at age 7 is the sum of the variances of F1 and F2, and that at age 9 is the sum of the variance of F1 and the variance of F2 multiplied by b_3^2, with a similar expression involving b_4 for age 11. Thus b_3 and b_4 are clearly involved in predicting both means and variances, the increases in the mean value being in proportion to the bs and the increases in the variance being in proportion to the squares of the bs.

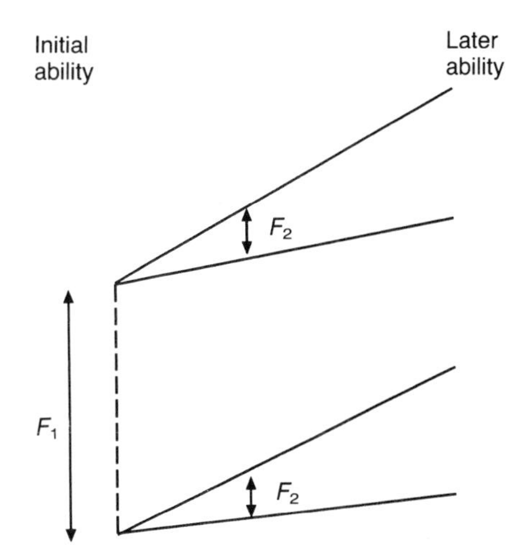

DISPLAY 6.29 Diagrammatic representation of ability trajectories implied by growth curve model.

Covariation in ability scores over time is also a necessary consequence of the structure of the model, since we are predicting four scores from two common factors. As the *b* coefficients increase so the proportion of variance due to initial score will decline. Thus if the *b* coefficients increase with time then the correlations with the initial scores are likely to show the desired pattern of decline.

The EQS command file for this model is shown in Display 6.30. In this model the four observed variables are not directly related to V999, but have their means determined solely through the factors. The factors F1 and F2 have their means determined by the coefficients on V999 and their variances by the variance of their corresponding disturbances D1 and D2. The starting values for the coefficients are simply the mean of the age 6 scores and the difference in means between age 6 and age 7. The starting value for the variance of D1 is a little smaller than the variance of the age 6 scores, and that for D2 is the difference in the variance of the scores at 6 and 7. We have assumed, as we have many times before, that the variances of the measurement errors are constant. The /PRINT EFFECT=YES; subcommand is included as EQS displays test statistics about the estimated means in the effect decomposition output. The second print subcommand COVARIANCE=YES; is also useful, displaying the predicted covariances and means for all observed and latent variables.

```
/TITLE
 Dependence between ability scores at 6, 7, 9 and 11
 Growth curve model of latent ability
/SPECIFICATIONS
 CASES=204; VARS=4; ANALYSIS=MOMENT; MATRIX=CORR;
/LABELS
 V1=ABIL6; V2=ABIL7; V3=ABIL9; V4=ABIL11;
/EQUATIONS
 V1=F1+E1;
 V2=F1+F2+E2;
 V3=F1+3*F2+E3;
 V4=F1+5*F2+E4;
 F1=18*V999+D1;
 F2=7*V999+D2;
/VARIANCES
 D1=25*; D2=1*;
 E1 TO E4=8*;
/CONSTRAINTS
 (E1,E1)=(E2,E2)=(E3,E3)=(E4,E4);
/PRINT
 COVARIANCE=YES;
 EFFECT=YES;
/MEANS
 18.034 25.819 35.255 46.593
/STANDARD DEVIATIONS
 6.374 7.319 7.796 10.386
/MATRIX
 1
 0.809    1
 0.806    0.850    1
 0.765    0.831    0.867    1
/END
```

DISPLAY 6.30 Growth curve model of latent ability.

The model fits with a 7 d.f. goodness-of-fit chi-squared statistic of 47.29 and a CFI of 0.951. The predicted covariance matrix and effect decomposition output is shown in Display 6.31. Other output only differs in the inclusion of an extra line for variable means in the output observed covariance matrix and in residuals between variables V1 to V4 and 'variable' V999. These are concerned with the difference between observed and predicted means.

The assumption that initial scores and trajectories are independent is not especially plausible. The circumstances that lead to a child having some initially high or low score, such as family

		V1	V2	V3	V4	V999	F1	F2
ABIL6	V1	44.866						
ABIL7	V2	36.419	47.378					
ABIL9	V3	36.419	41.914	56.883				
ABIL11	V4	36.419	45.604	56.509	78.450			
V999	V999	18.042	25.865	35.151	46.643	1.000		
F1	F1	36.419	36.419	36.419	36.419	18.042	36.419	
F2	F2	.000	2.512	5.495	9.185	7.822	.000	2.512

DECOMPOSITION OF EFFECTS WITH NONSTANDARDIZED VALUES

PARAMETER INDIRECT EFFECTS

```
ABIL6  =V1 = 18.042 V999 + 1.000 D1
              0.470
             38.417
ABIL7  =V2 = 25.865 V999 + 1.000 D1 +  1.000 D2
              0.482
             53.635
ABIL9  =V3 = 35.151 V999 + 1.000 D1 +  2.187 D2
               .529                     .069
             66.505                   31.739
ABIL11 =V4 = 46.643 V999 + 1.000 D1 +  3.656 D2
              0.621                     .119
             75.059                   30.755
```

DISPLAY 6.31 Model covariance matrix for measured and latent variables.

background, are likely to continue to operate over time. Thus high (low) initial ability and steeper (shallower) progress are likely to go together. A positive correlation between F1 and F2 is therefore to be expected. Fitting this model is left as an excercise to the reader.

The growth model perhaps provides a more natural framework in which to test hypotheses about uniform growth in ability than the autoregressive model. In particular, it is especially well suited to circumstances where the measures have not been taken at equally spaced intervals. In the autoregressive model we had to use the device of 'phantom' factors. For the growth model, we simply restrict the coefficients b_3 and b_4 (in fact their ratios to b_2, but that has been fixed at 1) to be in proportion to their time since T1 (where the interval between T1 and T2 is taken as 1). With intervals from T1 to T2 of 1 year, T1 to T3 of 3 years and T1 to T4 of

5 years, this suggests constraining b_3 to 3 and b_4 to 5. This model results in a far worse fit, suggesting such a model of uniformity to be wholly inapplicable. Exercise 6.6 explores a possible explanation for this.

Exercise 6.5

For the growth curve example of Display 6.30 insert the additional line necessary to allow the two disturbance terms D1 and D2 to be correlated (a sensible starting value is 0). Does this significantly improve the fit of the model?

Exercise 6.6

One reason why the ability data might not show uniform linear growth is that there are practice effects in the testing; in particular, the initial test might give relatively low scores in comparison to the later tests because it was new to the children. The addition of a direct link from V999 to V1 would allow for such an effect. Does the linear growth model fit better when allowing for such a practice effect?

6.12 SUMMARY

Although this chapter has focused on models for longitudinal data, many of the ideas and modelling techniques are of more general relevance. The decomposition of effects, into those that are direct and those that are indirect, is one of great value in the determination of likely mechanisms generating particular associations between variables. Although it is more persuasive in the analysis of longitudinal data, it is also frequently used for cross-sectional analyses. The lesson that the presence of measurement error can obscure the presence of a very simple underlying model is of almost universal relevance. The fact that, as here, the same data can be frequently analysed by using a variety of models, is also the case more generally. The investigator should not simply select the first basic model structure that comes to mind, then add and subtract bits until a 'well-fitting' model is found. The full range of alternatives should be considered, for both their theoretical and their empirical appropriateness.

Finally, two additional modelling techniques, involving the use of simultaneous modelling of means and covariances and the use of 'phantom' factors, have been illustrated. Considerable use of these two techniques is made in the following two chapters, greatly extending the range of data sets that can be analysed using EQS.

Simultaneous analysis of two or more groups

7

.1 INTRODUCTION

n the applications of structural modelling considered thus far, it as been assumed that the individuals supplying the data from vhich the observed correlation or covariance matrix is calculated re essentially a random sample from a single population. This ssumption may not always be reasonable, and many research nvestigations will involve individuals who can be identified as elonging to particular groups; examples include the gender of the ndividual, age group, diagnostic category, country of origin, and thnic community. For such data it may be of considerable interest o determine to what extent the covariance or correlation matrices ave the same structure in terms of some latent variable model. An investigator might, for example, try to discover whether some onfirmatory factor model for a set of cognitive tests is equally uitable for men and women. If exactly the same model is found ot to be appropriate for both sexes, then attempts may be made o allow some terms to differ to achieve a reasonable fit. Factor oadings, for example, may be constrained to be equal in the two roups but error variances allowed to differ.

A multigroup model will lead to a predicted value for the covari- nce or correlation matrix of each group, and again the parameters n the model are estimated by making these predicted values as close' as possible to the observed values. Here the overall measure of 'closeness' will be simply a weighted average of the closeness values of individual groups, found from one or other of the criteria lescribed in Chapter 3. It is important to note that multigroup nodels are, in general, harder to estimate than those involving a single group. Good starting values are often critical to obtaining well-behaved convergence. The greater the number of groups, the more difficult it will be to find an acceptable fit to the data from all the groups simultaneously.

7.2 TESTING THE EQUALITY OF CORRELATION AND COVARIANCE MATRICES

The initial stage in a multigroup analysis should be to assess whether the groups under investigation might have the same population covariance or correlation matrix. If this is the case, the different sample covariance/correlation matrices would simply be estimates of the same single-population matrix. Structural models evaluated on data from the different groups would be describing the same population, and so the models should be identical apart from chance variations. In such cases, rather than estimating the parameters of the model from each group, a multigroup analysis analysing all data simultaneously should be used, the model being specified to be identical in each group.

Where the covariance/correlation matrices of the groups are found to differ, models in which some parameters are allowed to differ might be evaluated in an attempt to account for the observed discrepancies. Examples will be given later in this chapter.

Testing the hypothesis that the covariance or correlation matrices of the different groups are equal can be done using EQS, although the resulting programs are rather lengthy. The basic approach is described in Display 7.1. As can be seen, it becomes necessary to

> For each observed variable the appropriate EQS program will contain an equation of the form
>
> ```
> V1=*F1;
> ```
>
> without an error term. The variance of each F variable is fixed at unity, but the covariances of each pair of F variables are free parameters. To test the equality of the correlation matrices, these covariances are constrained to be equal across groups. To test the equality of the covariances matrices, in addition to the latent variable covariances, the loadings of each observable variable on each latent variable must be fixed to be the same in each group.

DISPLAY 7.1 Testing the equality of covariance and correlation matrices in EQS.

Cross-group equality constraints must be specified in the last group, using the /CONSTRAINTS paragraph, and the following format

```
/CONSTRAINTS
(1,P1)=(2,P1)=(3,P1);
```

where 1, 2 and 3 refer to groups and P1 is any free parameter, designated in practice by a specific double-label name. Specific examples might be

```
/CONSTRAINTS
(1,E1,E1)=(2,E1,E1);   !the variance of E1 is equal in groups 1 and 2
(1,F1,F2)=(2,F1,F2);   !the variance of F2 is equal in groups 1 and 2
```

DISPLAY 7.2 Constraining parameters in EQS to be equal across groups.

(a) Good readers, n=75

	Variable									
Var.	1	2	3	4	5	6	7	8	9	10
1	6.92									
2	2.75	6.55								
3	2.23	1.86	6.50							
4	1.62	1.55	1.88	5.20						
5	2.45	2.23	1.77	1.14	3.72					
6	−0.28	0.78	1.24	1.31	0.85	4.84				
7	0.63	1.36	1.24	0.99	1.06	2.27	7.02			
8	−0.64	−0.34	0.59	0.38	0.78	1.70	2.41	6.00		
9	1.07	0.20	1.67	1.50	1.34	0.23	1.00	2.55	8.76	
10	0.63	0.97	2.36	1.96	1.09	1.32	2.81	2.38	2.20	5.06

(b) Poor readers, n=75

	Variable									
Var.	1	2	3	4	5	6	7	8	9	10
1	9.06									
2	6.12	10.05								
3	4.76	4.43	5.71							
4	3.90	4.11	2.42	5.62						
5	5.36	6.10	3.88	3.06	7.95					
6	3.05	2.01	2.12	2.45	1.27	6.97				
7	4.07	3.86	3.28	2.40	3.18	2.53	5.43			
8	4.08	3.28	2.42	1.59	3.52	1.61	3.86	8.70		
9	3.54	2.45	2.96	1.69	3.08	0.82	1.64	3.69	9.55	
10	3.43	4.29	3.13	2.05	2.83	3.06	3.17	4.70	2.97	5.95

DISPLAY 7.3 Covariance matrices for good and poor readers.

fix some parameters to be equal across groups. This is achieved in EQS as explained in Display 7.2.

To illustrate the testing procedure, data collected by Yule *et al* (1969) will be used. These data consist of scores on each of ten cognitive tests from the Wechsler series, administered to 150 children. By means of their scores on a test of reading ability, the children were divided into 'good' readers and 'poor' readers. The covariance matrices of the two groups are shown in Display 7.3.

```
/TITLE
 Test for equality of covariance matrices
 of good and poor readers
 Group 1 – good readers
/SPECIFICATIONS
 GROUPS=2; CASES= 75; VARIABLES=10; METHOD=ML;
 MATRIX=COV; ANALYSIS=COV;
! The groups parameter specifies the number of groups
/EQUATIONS
 V1=2*F1;
 V2=2*F2;
 V3=2*F3;
 V4=2*F4;
 V5=2*F5;
 V6=2*F6;
 V7=2*F7;
 V8=2*F8;
 V9=2*F9;
 V10=2*F10;
! Note that there are no error terms in these equations
/VARIANCES
 F1 TO F10=1.0;
! Variances of F variables are set to one
/COVARIANCES
 F1 TO F10=0.2*;
! Covariances of F variables are free parameters
! Shorthand notation is used
/MATRIX

COVARIANCE MATRIX FOR GOOD READERS

/END
```

DISPLAY 7.4 EQS program for testing equality of covariance matrices for good and poor readers.

```
/TITLE
Group 2 – poor readers
! Title information is optional
/SPECIFICATIONS
CASES=75; VARIABLES=10; METHOD=ML; MATRIX=COV;
ANALYSIS=COV;
! Note that groups parameter not needed
/EQUATIONS
 V1=2*F1;
 V2=2*F2;
 V3=2*F3;
 V4=2*F4;
 V5=2*F5;
 V6=2*F6;
 V7=2*F7;
 V8=2*F8;
 V9=2*F9;
 V10=2*F10;
/VARIANCES
 F1 TO F10=1.0;
/COVARIANCES
 F1 TO F10=0.2*;
/CONSTRAINTS
 (1,F1,F2)=(2,F1,F2);   ! Across group constraints
 (1,F1,F3)=(2,F1,F3);   !Each of the 45 covariances
                        ! are constrained to be equal in the two groups.
                        ! To save space not all the constraints are listed
                        here

(1,F9,F10)=(2,F9,F10);      ! 45 lines for covariance constraints
(1,V1,F1)=(2,V1,F1);        ! Across group constraints on
                            ! loadings
(1,V2,F2)=(2,V2,F2);

down to

(1,V10,F10)=(2,V10,F10);
/MATRIX

COVARIANCE MATRIX FOR POOR READERS

/END
```

DISPLAY 7.4 (*Continued*)

The EQS program which gives a test of the equality of the two matrices is listed in Display 7.4. The chi-squared goodness-of-fit statistic takes the value 93.05 with 55 degrees of freedom. The corresponding p-value is 0.001, and it is clear that the hypothesis that the two covariance matrices are identical should be rejected.

7.3 TESTING A COMMON FACTOR STRUCTURE FOR GOOD AND POOR READERS

Having shown that the covariance matrices of the good and poor readers are not the same, it becomes of interest to investigate models which might explain the difference. Here prior information about the tests suggests that the first five are 'verbal' and the last five 'performance' tests. Consequently a two-factor model might be considered to account for the covariances of the observed variables. A multigroup EQS program which fits such a model to each group without any constraints is shown in Display 7.5. The goodness-of-fit information given in Display 7.6 suggests that the fit of the model is less than satisfactory. Nevertheless, the example can still be used to illustrate the fitting of more restricted models arising from imposing across-group constraints. (Readers should note that the model fitted by the program in Display 7.5 is, of course, exactly equivalent to fitting the two-factor model *separately* to good and poor readers. The chi-squared statistic of 115.795 with 68 d.f. is simply the sum of the separate chi-squared values, 68.509, with 34 d.f., for good readers, and 47.286, with 34 d.f., for poor readers).

An EQS program for fitting an *identical* two-factor model in each group is shown in Display 7.7, and one in which only the factor loadings in the two groups are constrained to be equal is given in Display 7.8. A summary of the goodness-of-fit indicators for each model is given in Display 7.9. Comparing the fit of these constrained models with each other and with the fit of the unconstrained two factor model (see Display 7.6) shows that constraining the factor loadings to be equal in each group does not reduce the fit significantly, but similar constraints on the factor correlation and the error variances do reduce the fit.

```
/TITLE
Unconstrained two factor model
/SPECIFICATIONS
 GROUPS=2; CASES=75; VARIABLES=10; METHOD=ML;
 MATRIX=COV; ANALYSIS=COV;
/EQUATIONS
 V1=1*F1+E1;   ! F1 is verbal factor
 V2=1*F1+E2;
 V3=1*F1+E3;
 V4=1*F1+E4;
 V5=1*F1+E5;
 V6=1*F2+E6;   ! F2 is performance factor
 V7=1*F2+E7;
 V8=1*F2+E8;
 V9=1*F2+E9;
 V10=1*F2+E10;
/VARIANCES
 F1 TO F2=1.0;
 E1 TO E10=4.0*;
/COVARIANCES
 F1,F2=0.3*;
/MATRIX

 COVARIANCE MATRIX FOR GOOD READERS

/END
/SPECIFICATIONS
 CASES=75; VARIABLES=10; METHOD=ML; MATRIX=COV;
 ANALYSIS=COV;
/EQUATIONS

 As for Group 1

/VARIANCES
 F1 TO F2=1.0;
 E1 TO E10=4.0*;
/COVARIANCES
 F1,F2=0.3*;
/MATRIX

 COVARIANCE MATRIX FOR POOR READERS

/END
```

DISPLAY 7.5 EQS program for fitting an unconstrained two-factor model to good and poor reader covariance matrices.

(1) Chi-squared 115.795, d.f.=68, $p<0.001$
(2) AIC = −20.205
(3) Fit indices:
  (a) NFI 0.797
  (b) NNFI 0.868
  (c) CFI 0.900

Note that the chi-squared statistic is highly significant and fit indices are less than 0.9. This indicates that the model provides a very poor fit to the data.

DISPLAY 7.6 Goodness-of-fit summary for model fitted by EQS program in Display 7.5.

7.4 A MULTIGROUP MODEL FOR LIBERAL–CONSERVATIVE ATTITUDES AT THREE TIME POINTS

Judd and Milburn (1980) used a latent variable analysis to examine attitudes in a nation-wide sample of individuals who were surveyed on three occasions, in 1972, 1974 and 1976. Part of the data involved recording attitudes on three topics:

busing – a policy designed to achieve school integration;
criminals – the protection of the legal rights of those accused of crimes;
jobs – whether government should guarantee jobs and standard of living.

The sample consisted of 143 individuals each with four years of college education, and 203 individuals who had no college education. Display 7.10 shows the observed correlation matrices and standard deviations for the two groups. Judd and Milburn postulated that the interrelationships among the attitude measurements would be accounted for largely by a general factor of *liberalism–conservatism*, to which all three of the observed variables would be related at each of the three time periods, plus a specific factor for each attitude that would persist across time. They also assumed that liberalism in 1974 would be partly predictable from liberalism in 1972, and that, similarly, liberalism in 1974 would be predictive of liberalism in 1976. The path diagram for such a model is shown in Display 7.11. Fitting the model separately to each group gives chi-squared values of 11.65 (college), and 12.91 (no college), each with 16 degrees of freedom. The model appears to fit well in each group. If a multigroup analysis is carried out constraining the

```
/TITLE
Identical factor models for good and poor readers
/SPECIFICATIONS
 GROUPS=2; CASES=75; VARIABLES=10; METHOD=ML;
 MATRIX=COV; ANALYSIS=COV;
/EQUATIONS
 V1=1*F1+E1;
 V2=1*F1+E2;
 V3=1*F1+E3;
 V4=1*F1+E4;
 V5=1*F1+E5;
 V6=1*F2+E6;
 V7=1*F2+E7;
 V8=1*F2+E8;
 V9=1*F2+E9;
 V10=1*F2+E10;
/VARIANCES
 F1 to F2=1.0;
 E1 to E10=0.5*;
/COVARIANCES
 F1,F2=0.3*;
/MATRIX

 COVARIANCE MATRIX FOR GOOD READERS HERE

/END
/SPECIFICATIONS
 CASES=75; VARIABLES=10; METHOD=ML; MATRIX=COV;
 ANALYSIS=COV;
/EQUATIONS

 As in Group 1, to

/COVARIANCES
 F1,F2=0.3*;
```

DISPLAY 7.7 EQS program for fitting identical two-factor models to covariance matrices of good and poor readers.

```
/CONSTRAINTS
 (1,V1,F1)=(2,V1,F1);   ! Constrain loadings on F1
 (1,V2,F1)=(2,V2,F1);
 (1,V3,F1)=(2,V3,F1);
 (1,V4,F1)=(2,V4,F1);
 (1,V5,F1)=(2,V5,F1);
 (1,V6,F2)=(2,V6,F2);   ! Constrain loadings on F2
 (1,V7,F2)=(2,V7,F2);
 (1,V8,F2)=(2,V8,F2);
 (1,V9,F2)=(2,V9,F2);
 (1,V10,F2)=(2,V10,F2);
 (1,F1,F2)=(2,F1,F2);   ! Constrain factor correlation
 (1,E1,E1)=(2,E1,E1);   ! Constrain error variances
 (1,E2,E2)=(2,E2,E2);
 (1,E3,E3)=(2,E3,E3);
 (1,E4,E4)=(2,E4,E4);
 (1,E5,E5)=(2,E5,E5);
 (1,E6,E6)=(2,E6,E6);
 (1,E7,E7)=(2,E7,E7);
 (1,E8,E8)=(2,E8,E8);
 (1,E9,E9)=(2,E9,E9);
 (1,E10,E10)=(2,E10,E10);
  ! N.B. Parameters which are constrained to be equal across
  ! groups must be given the same starting values
/MATRIX

COVARIANCE MATRIX FOR POOR READERS HERE

/END
```

DISPLAY 7.7 (*Continued*)

The constraints now needed are:

```
/CONSTRAINTS
 (1,V1,F1)=(2,V1,F1);
        ⋮
 (1,V10,F2) = (2,V10,F2);
```

Factor correlations and error variances are now left unconstrained across groups.

DISPLAY 7.8 Changes to program in Display 7.7 needed to fit the two-factor model to good and poor reader covariance matrices in which only the factor loadings are constrained to be equal across groups.

	Identical models in both groups	Only loadings equal in both groups
(1) Chi-squared	150.427	126.650
d.f.	89	78
p	<0.001	<0.001
(2) AIC	−27.573	−29.35
(3) Fit indices:		
NFI	0.736	0.778
NNFI	0.871	0.883
CFI	0.872	0.899

Both models fit poorly.

DISPLAY 7.9 Goodness-of-fit summary for models fitted by programs in Displays 7.7 and 7.8.

model to be identical in the two groups, the resulting chi-squared value is 153.98 with 61 d.f. The associated *p*-value is less than 0.001, and the model clearly does not provide an acceptable fit.

Display 7.12 lists an EQS program for an intermediate model. Here factor loadings are constrained to be equal across groups, as are the regression coefficients of the liberalism latent variable for 1976 and 1974 on the previous occasion. The chi-squared statistic now takes the value 45.67, a reduction of 108 from the previous model, with the degrees of freedom being reduced by 20. Clearly this represents a substantially improved fit. The new model appears to fit the data very well, and a number of the parameter estimates are given in Display 7.13. It may be of interest, however, to determine whether releasing any of the current constraints would lead to an improvement in fit, and this can be investigated using the /LMTEST paragraph, placed before the final /END. In a multigroup analysis this test assesses the cross-group constraints. The results from applying the test in this case are shown in Display 7.14. It appears that releasing the eighth constraint, that involving the equality of the regression coefficient of liberalism in 1974 on liberalism in 1972, may improve the fit. Fitting such a model gives a chi-squared value of 39.616 with 40 degrees of freedom. This represents a small but statistically significant improvement in fit. (The significance levels of these tests should not be interpreted too

(a) College education $n = 143$

		1972			1974			1976		
		B	C	J	B	C	J	B	C	J
1972	B	1.00								
	C	0.43	1.00							
	J	0.47	0.29	1.00						
1974	B	0.79	0.43	0.48	1.00					
	C	0.39	0.54	0.38	0.45	1.00				
	J	0.50	0.28	0.56	0.56	0.35	1.00			
1976	B	0.71	0.37	0.49	0.78	0.44	0.59	1.00		
	C	0.27	0.53	0.18	0.35	0.60	0.20	0.34	1.00	
	J	0.47	0.29	0.49	0.48	0.32	0.61	0.53	0.28	1.00
SD		2.03	1.84	1.67	1.76	1.68	1.48	1.74	1.83	1.54

(b) No college education $n = 203$

		1972			1974			1976		
		B	C	J	B	C	J	B	C	J
1972	B	1.00								
	C	0.24	1.00							
	J	0.39	0.25	1.00						
1974	B	0.44	0.22	0.22	1.00					
	C	0.20	0.53	0.16	0.25	1.00				
	J	0.31	0.21	0.52	0.30	0.21	1.00			
1976	B	0.54	0.21	0.22	0.58	0.25	0.21	1.00		
	C	0.14	0.40	0.13	0.13	0.44	0.23	0.17	1.00	
	J	0.30	0.25	0.48	0.33	0.16	0.41	0.28	0.14	1.00
SD		1.25	2.11	1.90	1.31	1.97	1.82	1.34	2.00	1.79

KEY:

B Busing
C Criminals
J Jobs

DISPLAY 7.10 Correlations and standard deviations for attitudes at three time points.

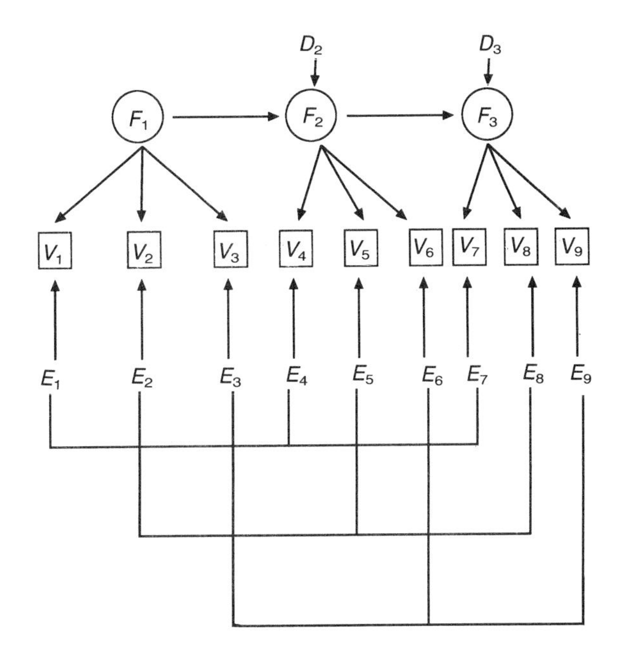

DISPLAY 7.11 Path diagram for attitudes in 1972, 1974 and 1976.

rigidly because of the problem of multiple testing.) For the college-educated sample, the regression coefficient of liberalism in 1974 on that in 1972 is estimated to be 1.118, with standard error 0.136; for those without a college education the estimate is 0.801, with standard error 0.109.

7.5 A MULTIGROUP EXAMPLE FROM GENETICS

The following example involves an investigation of the genetics of numerical ability, described in Loehlin and Vandenberg (1968). The data involve three subscales of the Number factor in Thurstone's Primary Mental Abilities battery, in male and female identical (Mz) and fraternal (Dz) twin pairs. The relevant correlation matrices and standard deviations for the four groups are shown in Display 7.15. In Display 7.16 a path diagram is shown which describes the assumed genetic influences on the covariances within and across twins. The latent variable here refers to a general genetic predisposition to do well on numerical tests. The loadings of each observed variable on the latent variable are assumed to be

```
/TITLE
liberalism-conservatism example
factor loadings and latent variable
regression coefficients constrained to be
equal across groups
group 1 – four years of college education
/SPECIFICATIONS
 GROUPS=2; CASES=143; VARIABLES=9;
 METHOD=ML; MATRIX=CORR; ANALYSIS=COV;
/EQUATIONS
 V1=1*F1+E1;     ! F1 is liberalism in 1972
 V2=1*F1+E2;
 V3=1*F1+E3;
 V4=F2+E4;       ! F2 is liberalism in 1974
                 ! Scale of F2 set to that of V4
                 ! Scale cannot be set in /VARIANCE
                 ! paragraph since F2 appears later
                 ! as a dependent variable
 V5=1*F2+E5;
 V6=1*F2+E6;
 V7=F3+E7;       ! F3 is liberalism in 1976
                 ! Again scale set to that of observed variable
 V8=1*F3+E8;
 V9=1*F3+E9;
 F2=1*F1+D1;     ! Regression of 1974 liberalism on 1972
 F3=1*F2+D2;     ! Regression of 1976 liberalism on 1974
/VARIANCES
 F1=1.0;
 E1 TO E9=1.0*;
 D1 TO D2=0.2*;
/COVARIANCES
 E1,E4=0.5*;
 E1,E7=0.5*;
 E2,E5=0.5*;
 E2,E8=0.5*;
 E3,E6=0.5*;
 E3,E9=0.5*;
 E4,E7=0.5*;
 E5,E8=0.5*;
 E6,E9=0.5*;
/MATRIX

Correlation matrix for college educated
```

DISPLAY 7.12 EQS program for constrained multigroup analysis of liberalism among college and non-college individuals.

```
/STANDARD DEVIATIONS

Standard deviations for college educated

/END
/TITLE
Group 2 – no college education
/SPECIFICATIONS
 CASES=203; VARIABLES=9; METHOD=ML;
 MATRIX=CORR; ANALYSIS=COV;
/EQUATIONS

As for Group 1, to

/COVARIANCES

As for Group 1

/CONSTRAINTS
(1,V1,F1)=(2,V1,F1);   ! Constrain factor loadings
(1,V2,F1)=(2,V2,F1);
(1,V3,F1)=(2,V3,F1);
(1,V5,F2)=(2,V5,F2);
(1,V6,F2)=(2,V6,F2);
(1,V8,F3)=(2,V8,F3);
(1,V9,F3)=(2,V9,F3);
                       ! Note no constraint specified for V4 or V7
(1,F2,F1)=(2,F2,F1);   ! Constrain regression coefficient
                       ! 1974 on 1972
(1,F3,F2)=(2,F3,F2);   ! Constrain regression coefficient
                       ! 1976 on 1974
/MATRIX

Correlations for no college education

/STANDARD DEVIATIONS

Standard deviations for no college education

/END
```

DISPLAY 7.12 (*Continued*)

(1) Factor loadings

		Estimated loading	S.E.	z
1972	B	1.113	0.112	9.94
	C	0.840	0.120	7.009
	J	1.000	0.116	8.601
1974	B	Fixed at 1.0		
	C	0.771	0.130	5.923
	J	0.904	0.147	6.163
1976	B	Fixed at 1.0		
	C	0.557	0.124	4.495
	J	0.845	0.144	5.879

(2) Regression coefficients for latent variables

	Estimated Coefficient	S.E.	z
1974–1972	0.93	0.103	9.096
1976–1974	0.99	0.086	11.541

All loadings and regression coefficients significantly different from zero.

DISPLAY 7.13 Parameter estimates for model fitted by EQS program in Display 7.12.

the same for both twins of a pair, and the genetic predispositions are assumed to be perfectly correlated for Mz twins, but to have a correlation of 0.5 for Dz twins. The residual variances which are here attributable to non-genetic factors including measurement error, are also constrained to be the same for each twin of a pair. In addition to these *within-groups* constraints, models of interest will include *between-groups* constraints. A possible model, for example, might allow males and females to have different parameter values, but keep those for identical and fraternal twins the same. The complete EQS program for such a model is given in Display 7.17.

The standardized solutions for males and females are shown in Display 7.18. The genetics paths all have values of about 0.8. The squares of these values represent the proportion of variance due to genetic factors. So here (if the model is correct – see later) the variance in these measures of numerical ability is approximately

(1) List of constraints

```
CONSTR:1 (1,V1,F1) – (2,V1,F1) = 0;
CONSTR:2 (1,V2,F1) – (2,V2,F1) = 0;
CONSTR:3 (1,V3,F1) – (2,V3,F1) = 0;
CONSTR:4 (1,V5,F2) – (2,V5,F2) = 0;
CONSTR:5 (1,V6,F2) – (2,V6,F2) = 0;
CONSTR:6 (1,V8,F3) – (2,V8,F3) = 0;
CONSTR:7 (1,V9,F3) – (2,V9,F3) = 0;
CONSTR:8 (1,F2,F1) – (2,F2,F1) = 0;
CONSTR:9 (1,F3,F2) – (2,F3,F2) = 0;
```

(2) Univariate test statistics

Constraint	Chi-squared	*p*
1	2.154	0.142
2	0.004	0.951
3	0.192	0.661
4	0.195	0.659
5	0.312	0.576
6	0.135	0.714
7	0.187	0.665
8	6.158	0.013
9	1.304	0.254

(3) Multivariate statistics

	Cumulative multivariate statistics			Univariate increment	
Constr.	Chi-sq.	d.f.	*p*	Chi-sq.	*p*
8	6.158	1	0.013	6.158	0.013
1	11.782	2	0.003	5.625	0.018
7	13.763	3	0.003	1.980	0.159
5	17.172	4	0.002	3.409	0.065
4	19.411	5	0.002	2.239	0.135
6	21.040	6	0.002	1.628	0.202
9	21.289	7	0.003	0.249	0.618
2	21.399	8	0.006	0.110	0.740
3	21.431	9	0.011	0.032	0.858

Releasing the constraint for the regression coefficient of liberalism in 1974 on that in 1972 to be the same in both groups should improve the fit of the model.

DISPLAY 7.14 Results of applying LMTEST to assess constraints in EQS program latent in Display 7.12.

(1) Male identical twins, $n=63$

	A1	M1	H1	A2	M2	H2
Addition 1	1.000					
Multiplication 1	0.670	1.000				
3-Higher 1	0.489	0.555	1.000			
Addition 2	0.598	0.499	0.526	1.000		
Multiplication 2	0.627	0.697	0.560	0.784	1.000	
3-Higher 2	0.456	0.567	0.725	0.576	0.540	1.000
S.D.	7.37	13.81	16.93	8.17	13.33	17.56

(2) Female identical twins, $n=59$

	A1	M1	H1	A2	M2	H2
Addition 1	1.000					
Multiplication 1	0.611	1.000				
3-Higher 1	0.754	0.676	1.000			
Addition 2	0.673	0.464	0.521	1.000		
Multiplication 2	0.622	0.786	0.635	0.599	1.000	
3-Higher 2	0.614	0.636	0.650	0.574	0.634	1.000
S.D.	8.00	12.37	15.19	6.85	11.78	14.76

(3) Male fraternal twins, $n=29$

	A1	M1	H1	A2	M2	H2
Addition 1	1.000					
Multiplication 1	0.664	1.000				
3-Higher 1	0.673	0.766	1.000			
Addition 2	0.073	0.313	0.239	1.000		
Multiplication 2	0.194	0.380	0.347	0.739	1.000	
3-Higher 2	0.379	0.361	0.545	0.645	0.751	1.000
S.D.	9.12	16.51	17.20	7.70	14.52	14.74

(4) Female fraternal twins, $n=46$

	A1	M1	H1	A2	M2	H2
Addition 1	1.000					
Multiplication 1	0.779	1.000				
3-Higher 1	0.674	0.679	1.000			
Addition 2	0.462	0.412	0.500	1.000		
Multiplication 2	0.562	0.537	0.636	0.620	1.000	
3-Higher 2	0.392	0.359	0.565	0.745	0.603	1.000
S.D.	8.99	15.44	16.98	7.65	14.59	18.56

DISPLAY 7.15 Within-individual and cross-pair correlations for three sub-tests of numerical ability in male and female, Mz and Dz twins (taken with permission from Loehlin, 1987).

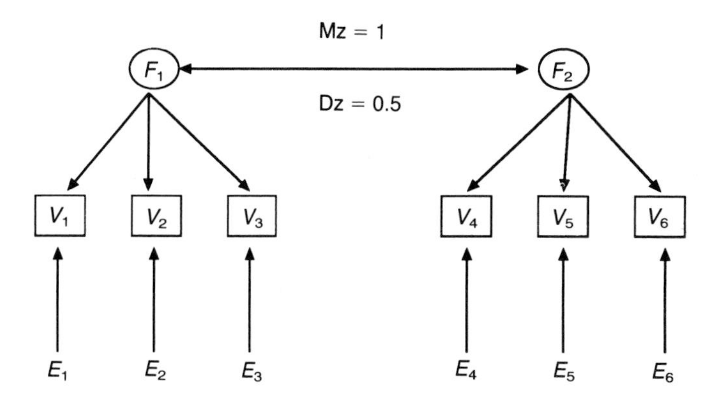

DISPLAY 7.16 Path diagram for twin scores.

60–65% genetic and 35–40% non-genetic. However, before taking such a conclusion at its face value, the fit of the model needs to be examined, and the necessary information is given in Display 7.19. Clearly the fit of the current model is not impressive. That this is the case is perhaps not very surprising, since the model as formulated does not allow for non-genetic contributions to the *correlations* between the observed variables. Including non-zero covariances between the error terms, constrained in particular ways, would overcome this problem – see Exercises 7.1 and 7.2.

Exercise 7.1

Use EQS to test the equality of the four correlation matrices in the genetic example.

Exercise 7.2

Refit the genetics data with a model which allows error terms to be correlated. Remember that because of the arbitrary labelling of the twins, several of the error correlation terms will need to be fixed to be equal to one another within each group.

7.6 STRUCTURED MEANS IN MULTIGROUP SAMPLES

The final example to be discussed in this chapter concerns the use of EQS to assess differences between group *means* on the latent

```
/TITLE
Genetics of numerical ability
Loehlin (1987)
Group 1 – identical male twins
/SPECIFICATIONS
 GROUPS=4; CASES=63; VARIABLES=6; METHOD=ML;
 MATRIX=CORR; ANALYSIS=COV;
/LABELS
 V1=A1; V2=M1; V3=H1; V4=A2; V5=M2; V6=H2;
/EQUATIONS
 V1=1*F1+E1;
 V2=1*F1+E2;
 V3=1*F1+E3;
 V4=1*F2+E4;
 V5=1*F2+E5;
 V6=1*F2+E6;
/VARIANCES
 F1=1.0;
 F2=1.0;             ! Factors in standardized form
 E1 TO E6=0.5*;
/COVARIANCES
 F1,F2=1.0;          ! Factor correlation set to unity for
                     ! identical twins
/CONSTRAINTS
 (E1,E1)=(E4,E4);    ! Constrain residual variances of each
 (E2,E2)=(E5,E5);    ! twin to be the same
 (E3,E3)=(E6,E6);
 (V1,F1)=(V4,F2);    ! Constrain factor loadings of twins
 (V2,F1)=(V5,F2);    ! to be the same
 (V3,F1)=(V6,F2);
/MATRIX

Correlations of male identical twins here

/STANDARD DEVIATIONS

Standard deviations of male identical twins here

/END
```

DISPLAY 7.17 EQS program for genetic example.

```
/TITLE
Group 2 – female identical twins
/SPECIFICATIONS
 CASES=59; VARIABLES=6; METHOD=ML; MATRIX=CORR;
 ANALYSIS=COV;

Commands as for Group 1, to

/MATRIX

Correlations for female identical twins

/STANDARD DEVIATIONS

Standard deviations for female identical twins

/END
/TITLE
Group 3 – male fraternal twins
/SPECIFICATIONS
 CASES=29; VARIABLES=6; METHOD=ML; MATRIX=CORR;
 ANALYSIS=COV;

Commands as in Group 1, to

/MATRIX

Correlations for male fraternal twins

/STANDARD DEVIATIONS

Standard deviations for male fraternal twins

/END
```

DISPLAY 7.17 (*Continued*)

variables of interest. The relevant procedures will involve the 'constant' variable V999 introduced in Section 3.6. Since the models to be fitted make predictions about the means of the observed variables in addition to predictions about their covariances, the 'closeness' criterion used will be that invoked by the use of ANALYSIS=MOMENT in the /SPECIFICATIONS section. In this way parameters of the model will be estimated to make both

```
/TITLE
Group 4 – female fraternal twins
/SPECIFICATIONS
 CASES=46; VARIABLES=6; METHOD=ML; MATRIX=CORR;
 ANALYSIS=COV;

Commands as in Group 1, to

/CONSTRAINTS
 (E1,E1)=(E4,E4);
 .                        ! Within group constraints as
 .                        ! in Group 1
 (V3,F1)=(V6,F2);
 (1,E1,E1)=(3,E1,E1);     ! Set error variances for male
(1,E2,E2)=(3,E2,E2);      ! identical and fraternal twins to be same
(1,E3,E3)=(3,E3,E3);
(1,V1,F1)=(3,V1,F1);      ! Set factor loadings for male
(1,V2,F1)=(3,V2,F1);      ! identical and fraternal twins to be same
(1,V3,F1)=(3,V3,F1);
(2,E1,E1)=(4,E1,E1);      ! Set error variances for female
(2,E2,E2)=(4,E2,E2);      ! identical and fraternal twins to be same
(2,E3,E3)=(4,E3,E3);
(2,V1,F1)=(4,V1,F1);      ! Set factor loadings for female
(2,V2,F1)=(4,V2,F1);      ! identical and fraternal twins to be same
(2,V3,F1)=(4,V3,F1);
/MATRIX

Correlations for female fraternal twins

/STANDARD DEVIATIONS

Standard deviations for female fraternal twins

/END
```

DISPLAY 7.17 (*Continued*)

the differences between predicted and observed means and the corresponding differences for covariances 'small'. In some cases, of course, fitting both means and covariances adequately may not be possible and, as a consequence, these **structured mean** models can be difficult to apply in practice, especially when many variables and parameters are involved.

The essentials of the models of interest here can be most usefully

(1) Males

Variable	Genetic path	Residual variance
Addition	0.774	0.634
Multiplication	0.846	0.532
3-Higher	0.766	0.643

(2) Females

Variable	Genetic path	Residual variance
Addition	0.803	0.597
Multiplication	0.825	0.564
3-Higher	0.807	0.590

DISPLAY 7.18 Standardized solutions for males and females from EQS program in Display 7.17.

(1) Chi-squared = 131.399, d.f.=72, $p<0.001$
(2) AIC = −12.601
(3) Fit indices:
NFI = 0.826
NNFI = 0.929
CFI = 0.915

The fit is not very impressive.

DISPLAY 7.19 Summary of goodness-of-fit measures for model fitted by EQS program in Display 7.17.

described by means of an example, and for this purpose part of the data collected by Kluegel, Singleton and Starnes (1977) will be used. These data consist of observations made on two different ethnic groups, the first containing 432 members and the second 368. Interest centred on possible differences between the groups on their perception of their own 'class status'. Four variables were recorded for each individual, each concerned with the individual's assessment of their perceived class status: occupation; income; life-style; and influence. For each of these variables respondents placed themselves in one of four class categories – 'lower', 'working', 'middle', or 'upper' – scored 0 to 3.

It was postulated that the four variables would be indicators of a single underlying latent variable, which will be labeled simply *sub-*

	Correlations			
	Occupation	Variable income	Lifestyle	Influence
Occupation	1.00	0.55	0.57	0.48
Income	0.39	1.00	0.65	0.52
Lifestyle	0.39	0.55	1.00	0.65
Influence	0.38	0.46	0.53	1.00

(Group 1 below diagonal, Group 2 above diagonal)

	Means and standard deviations			
	Occupation	Income	Lifestyle	Influence
Group 1				
Mean	1.54	1.55	1.54	1.60
S.D.	0.64	0.67	0.62	0.65
Group 2				
Mean	1.29	1.13	1.23	1.32
S.D.	0.75	0.81	0.79	0.86

DISPLAY 7.20 Correlations, means and standard deviations for indicators of subjective class in two ethnic groups.

jective class, and the primary goal was to estimate the difference in the mean of this variable in the two groups. The correlations, means and standard deviations of the four variables in each of the two groups are given in Display 7.20. The first task is to assess whether the suggested single-factor model does hold in both groups. The results from fitting such a model, first allowing all parameters to differ and then constraining loadings to be the same in each group, are shown in Display 7.21. Clearly the constrained model provides a very good fit for the observed covariances, and this model may now be used in the assessment of the difference in the mean of the latent variable in the two groups. The latent variable means in each group will be given by the intercept parameters found using V999. But if an attempt is made to estimate *separate* intercept terms in the two groups a problem will be encountered. The difficulty arises because the origin of the latent variable is arbitrary. Fortunately, this is easily overcome by setting the intercept in one of the groups to zero. The estimated intercept in the other group then represents a *difference* in the means of the two groups. Finally, the intercept terms for the observed variables are constrained to be the same in each group, so that any differences in the means of these variables are accounted for by the difference

(1) Unconstrained model
(a) Chi-squared = 7.994, d.f. = 4, $p = 0.0918$
(b) AIC = −0.00632
(c) Fit indices:
NFI = 0.992
NNFI = 0.988
CFI = 0.996

(2) Factor loadings constrained to be equal in the two groups
(a) Chi-squared = 9.475, d.f. = 7, $p = 0.2203$
(b) AIC = −4.525
(c) Fit indices:
NFI = 0.991
NNFI = 0.996
CFI = 0.998

Note that the AIC criterion is lower for the constrained model. Also the increase in χ^2 from the unconstrained to the constrained model is very small.

DISPLAY 7.21 Comparison of fit of two single-factor models for the data in Display 7.20.

in the factor mean. A suitable EQS program for fitting the structured mean model to the data in Display 7.20 is given in Display 7.22. A summary of the output from this program is shown in Display 7.23. The model appears to fit quite well (see Display 7.24). The estimated difference in mean class in the two groups is 0.255, with the second group having the lower value. An approximate 95% confidence interval for this difference is $0.255 \pm 2 \times 0.036 = (0.183, 0.327)$.

Exercise 7.3

Compare the fit of the structured mean model fitted in the text with one which constrains the latent variable, subjective class to have the same mean in each ethnic group.

Exercise 7.4

Use the LMTEST on the structured mean model specified by the program in Display 7.22 to investigate how to improve the fit of the model.

```
/TITLE
Analysis of subjective class in two groups
/SPECIFICATIONS
 CASES=432; GROUPS=2; VARIABLES=4; METHOD=ML;
 MATRIX=CORR;
 ANALYSIS=MOMENT;          ! This indicates that means are
                           ! included in the model. Note also that
                           ! ME=ML is the only option allowed
                           ! with structured means
/LABELS
 V1=Occup; V2=Income; V3=Lifestyle; V4=Influence;
 F1=Subj-class;
/EQUATIONS
 V1=1.0*V999+1.0F1+E1;     ! Intercept to be estimated. Scale of F1
                           ! set to that of V1
 V2=1.0*V999+1.0*F1+E2;
 V3=1.0*V999+1.0*F1+E3;
 V4=1.0*V999+1.0*F1+E4;
 F1=0V999+D1;              ! Intercept (i.e. mean) of F1 set to zero
                           ! in this group. An equivalent form is
                           ! F1=D1;
/VARIANCES
 E1 TO E4=0.5*
 D1=0.5*;
/MATRIX

Correlation for Group 1 here

/MEANS

Means for Group 1 here

/STANDARD DEVIATIONS

Standard deviations for Group 1 here

/END
```

DISPLAY 7.22 EQS program to fit structured mean model to data from two ethnic groups.

```
/TITLE
Group 2
/SPECIFICATIONS
 CASES=368; VARIABLES=4; METHOD=ML; MATRIX=CORR;
 ANALYSIS=MOMENT;
/LABELS
 V1=Occup; V2=Income; V3=Lifestyle; V4=Influence;
 F1=Subj-class;
/EQUATIONS

Those involving observed variables as for group 1

F1=1.0*V999 + D1;   ! Intercept (i.e. mean) of F1 free to
                    ! be estimated in this group
/VARIANCES
As for Group 1, to

/MATRIX

Correlations for Group 2

/MEANS

Means for Group 2

/STANDARD DEVIATIONS

Standard deviations for Group 2

/CONSTRAINTS
 (1,V1,V999)=(2,V1,V999);   ! Intercepts of observed variables
                            ! constrained to be the same in each group
 (1,V2,V999)=(2,V2,V999);
 (1,V3,V999)=(2,V3,V999);
 (1,V4,V999)=(2,V4,V999);
 (1,V2,F1)=(2,V2,F1);       ! Factor loadings constrained to be
                            ! equal in each group
 (1,V3,F1)=(2,V3,F1);
 (1,V4,F1)=(2,V4,F1);
/PRINT
 EFFECT=YES;                ! This asks EQS to print out direct and
                            ! indirect effects – see Display 7.23
/END
```

DISPLAY 7.22 (*Continued*)

7.7 SUMMARY

Many research investigations involve separate groups of individuals. For such studies it is often of interest to assess which aspects of particular models are the same in the different groups and which are not the same. In a confirmatory factor analysis model, for example, factor loadings may be constrained to be equal across groups but factor correlations and/or error variances may be allowed to differ. Although such models can be specified relatively easily in an EQS program they are frequently more difficult to fit than their single-group equivalents. The problem becomes more acute as the number of groups increases.

1. Measurement equations with standard errors and z values – same in both groups

```
V1 =V1 = 1.000F1  +     1.542*V999 +   1.000E1
                        0.027
                        57.052
V2 =V2 = 1.307*F1 +     1.514*V999 +   1.000E2
         0.079          0.030
         16.527         50.806
V3 =V3 = 1.349*F1 +     1.555*V999 +   1.000E3
         0.078          0.029
         17.334         54.554
V4 =V4 = 1.237*F1 +     1.612*V999 +   1.000E4
         0.078          0.029
         15.877         55.565
```

2. Construct equations with standard errors and t-statistics

Group 1

```
F1 =F1 = 1.000D1
```

Group 2

```
F1 =F1 = -0.255*V999 + 1.000D1
          0.036
         -7.085
```

Note that the estimated intercept parameter in Group 2 gives the estimate of the difference between the means of the twogroups on subjective class.

DISPLAY 7.23 Summary of output from EQS program in Display 7.22.

3. Parameter total effects

Group 1
```
V1 =V1 = 1.000F1  +    1.542*V999 + 1.000E1 + 1.000D1
V2 =V2 = 1.307*F1 +    1.514*V999 + 1.000E2 + 1.307D1
V3 =V3 = 1.349*F1 +    1.555*V999 + 1.000E3 + 1.349D1
V4 =V4 = 1.237*F1 +    1.612*V999 + 1.000E4 + 1.237D1
F1 =F1 = 1.000D1
```

Group 2
```
V1 =V1 = 1.000F1  +    1.287*V999 + 1.000E1 + 1.000D1
V2 =V2 = 1.307*F1 +    1.181*V999 + 1.000E2 + 1.307D1
V3 =V3 = 1.349*F1 +    1.212*V999 + 1.000E3 + 1.349D1
V4 =V4 = 1.237*F1 +    1.298*V999 + 1.000E4 + 1.237D1
F1 =F1 = -0.255*V999 + 1.000D1
```

The coefficients of V999 give the predicted means for the observed variables. Note that for group 1 these are the same as the estimated intercepts since in this group F1 is assumed to have mean zero. For group 2 the predicted means are a combination of intercept and factor mean – for example, the predicted mean for variable 2 in group 2 is simply

$$1.514 + 1.307 \times (-0.255) = 1.514 - 0.333 = 1.181$$

The -0.333 in this calculation is printed out by EQS as an *indirect* effect.

DISPLAY 7.23 (*Continued*)

(1) Chi-squared=18.641, d.f.=10, p=0.045
(2) AIC=−1.3593
(3) Fit indices:
NFI = 0.982
NNFI = 0.990
CFI = 0.991

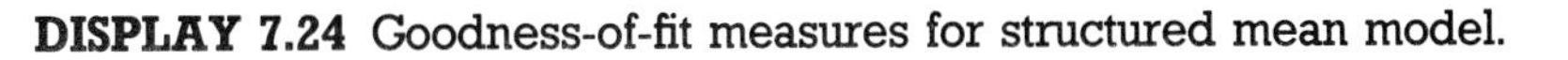

DISPLAY 7.24 Goodness-of-fit measures for structured mean model.

In examples with more than a single group it often becomes of interest to examine group mean differences on the latent variables. Again is this dealt with quite simply by EQS, but again the fitting procedure can be more difficult.

Specific points to note are the following:

1. The choice of starting values will be more critical for multigroup models.
2. When looking at differences of means of the latent variables it must be remembered that their origin is arbitrary. Consequently, the corresponding intercept parameter must be set equal to zero in one of the groups.

Common practical problems

8

8.1 INTRODUCTION

The previous chapters have introduced and explained various types of model and illustrated their use. The reader, in following through the examples, should have grasped the essential modelling concepts, terminology and techniques of the latent variable approach. However, in taking these skills and knowledge and attempting to apply them to new models and data, readers are likely to encounter a number of additional issues and problems that we have not yet discussed. This chapter deals with several issues, both practical and theoretical, that the researcher will need to confront when setting out to analyse his/her own study.

Section 8.2 is concerned with problems typically encountered in the actual model fitting on the computer: well-behaved examples are one thing, fitting models to unfamiliar data can be quite another. Section 8.3 is concerned with what to do with data that contain variables that are skewed or categorical. Such variables are not normally distributed as the standard assumptions of the method would require them to be. Section 8.4 discusses the all too common problem of missing data and explains and illustrates one method for dealing with it. The model fitting that has been done in the examples of this book has assumed, implicitly, that the data have been obtained by simple random sampling from some population. However, data are often collected by other more complicated sampling designs, and Section 8.5 discusses whether and how data from such designs can be analysed. Finally, Section 8.6 explains how hypothesis tests and comparisons of different models should be placed in the context of some knowledge as to the researcher's chances of being able to reject or distinguish them statistically. If the data are too scanty to be able to say whether one model fits better than another, then conclusions as to the simplicity of the population process may be quite misleading – sufficient data might show the presence of important additional complexity. The section explains how such questions of **statistical power** can be examined.

These issues often present problems rather more difficult than those covered in preceding chapters. We have tried to give clear guidance as to recommended procedures. However, sometimes we can suggest only partial or qualified solutions. At other times, though we may suggest one method, other authors or standard practice may suggest another. Although we have made clear the areas of such disagreement, an attempt at a thorough argument or critique would have been inappropriate for this book. The interested reader may wish to follow up the references.

8.2 MODEL FITTING IN PRACTICE

In practice, the efficient and effective fitting of latent variable models to real sets of data relies heavily on the use of skills and knowledge that are best described as statistical common sense. Although this is usually gained simply by experience, we thought it might be useful to make some suggestions as to what constitutes good practice.

Before trying to fit any sort of model, some basic familiarization with and checking of the data is always essential. Graphical procedures, such as histograms and scatter plots, and the computation of various summary descriptive statistics are the basic tools to use. Such procedures are valuable features of the Windows version of EQS. In addition to discovering possible outliers, miscoded data, duplicate cases, missing data and such like, visual inspection of the input data used in the model fitting is also essential. It is often time-consuming, and eventually pointless, to fit models when either the model is the wrong one or the data have little or no consistent structure. Much time can be saved by first checking that the model and data have some rough correspondence. EQS can often calculate the appropriate summary statistics required for any particular analysis (specified in the ANALYSIS part of the /SPECIFICATIONS paragraph) from a variety of forms of input data. Where possible (but see Sections 8.3 and 8.4), ve would strongly recommend using as input data the means, orrelations and standard deviations rather than raw data, covariance or moment matrices. This is because the former are much nore readily interpreted by eye. Trends in means and variances re immediately visible and the signs and magnitudes of correations can be compared with those expected under the model. It hould be emphasized that this is not a recommendation to have EQS actually analyse the correlation matrix, where analysing the covariance matrix might be more appropriate (commonly done but not recommended – see Cudeck, 1989). What data are input and what are analysed are related but distinct. EQS will, by default, calculate and analyse a covariance matrix even when the input data are standard deviations and correlations.

Once the input data are sorted out and the data look as if they have some chance of conforming to the model you wish to fit, do not be too ambitious. Start with something simple and add complications. Fitting a complicated model immediately is rarely successful and can quickly lead to frustration and disillusionment. There are a variety of simple ways to start:

1. Fit the variances and covariances first before proceeding to a joint model of both means and covariances.
2. In multigroup problems begin by fitting to each group separately.
3. With models involving many observed measures, begin with a suitable subset of the measurement equations, build these up progressively to the full set required, and then add any construct equations.
4. Fit the null hypothesis models, which typically contain restrictions on parameters (usually that their value is zero or equal to another parameter) or the omission of some parameters, before the more complicated alternatives. The estimated parameters from the solution of the null model can often, where required, be used as the starting values for fitting the alternatives.

Unfortunately, the suggestion to start with something simple, or to simplify the analysis when problems are encountered, can occasionally make things worse rather than better. For example, sometimes the omission of a relationship from a model results in, say, a failure of the program to converge, or an error variance being estimated as zero. With the inclusion of the relationship everything behaves well. Simplifying, by reducing the number of observed variables being analysed, also runs the risk of generating problems of identification. Some techniques for obtaining identifiability of common parts of models, such as sets of measurement equations, have been explained, but an exhaustive check of the identifiability of every model parameter can be difficult. When a model is simplified, the analyst should check that, at a minimum, the number of observed covariances (and means where they are being modelled) is greater than the number of parameters being estimated.

Sadly, the analyst cannot afford to relax even when the program successfully converges on some solution, even where the message 'NO SPECIAL PROBLEMS ENCOUNTERED' appears in the output. Firstly, EQS fits the program you specified, not the one you wanted to specify. The input lines that EQS prints as the first part of the output must be checked line by line. Are all the variables in the equations for which variances are required listed in the variances paragraph? Are all the equation coefficients that are to be estimated, and none of those that are not to be estimated, followed by asterisks? Are *all* the constraints required properly listed? Do all parameters that are constrained to be equal share the same starting values? These are just some of the checks that must be made to the input. In doing these checks it soon becomes clear that there is much to be gained by a completely systematic organization of the equations, variances, covariances and constraints. Once the input has been checked, so also should the output. For example, check that all parameters that were to be constrained equal are indeed estimated to be the same (though remember that EQS imposes constraints in the unstandardized parameter space so they may only be the same in the unstandardized solution and may be different in the standardized solution). Finally, check that the degrees of freedom claimed for a model is equal to the number of observed means and covariances minus the number of estimated parameters. Of course, EQS usually does this calculation correctly. However, there are a number of circumstances – for example, when a parameter, constrained equal to another, has gone to a bound – where the EQS output may not be the one you want.

In some of our examples the structure of the model to be fitted to the data was almost obvious. As a result, there was only a limited and structured exploration of the fitted model, primarily for purposes of hypothesis testing. However, in other examples some empirical exploration and perhaps *ad hoc* modification of the model was required to obtain a reasonable fit. As a rule, the more alteration to the model that is made, particularly if it is made solely on the basis of empirical lack of fit (for example, to remove or reduce large residuals), the less faith one can have that conclusions are generalizable – the eventual model becomes increasingly a description of the particular sample obtained rather than a description of the population from which it was drawn. This problem can be minimized by a careful assessment of the available theory relevant to the problem and by ensuring that model definition and subsequent modification are theoretically motivated. There is a similar need for considerable preliminary thought in the sequence in which hypothesis tests are undertaken, firstly to ensure a structured set of independent tests, and secondly to maximize the meaningfulness of standard measures of significance level that ignore the problems of multiple tests. Do not let tests of central theoretical importance be lost among a long sequence of uninteresting ones.

A final general recommendation is not to rest content once you have found a model that fits the data. There may be several others that may do so almost equally well if not better. Sometimes these alternative models may imply a radically different interpretation of the data, but more often careful thought will reveal them as largely different perspectives on the same process, each giving its own insight.

8.3 NON-NORMAL DATA

The standard application of latent variable models assumes that all the latent and observed variables jointly conform to the multivariate normal (MVN) distribution. The only exception to this in the examples of this book has been where categorical variables have been dealt with by stratifying by category to form multiple groups and analysing the data assuming it to be MVN within each stratum. This does not, however, cover all but a small fraction of the circumstances in which non-normal data appear in practice. Indeed, in Section 5.3 a variable with just two categorical values appeared in an example, and there its special characteristics were ignored, it being treated as if it were a continuous variable. That is not a practice that we can recommend.

8.3.1 Continuous non-normal data

Raw data, such as symptom totals from questionnaire and checklist sources, often possess an inverted *J*-shaped distribution, with many very low scores and only a few observations, representing subjects with some kind of illness or problem, showing substantial scores. The most common suggestion is to transform the raw scores and to analyse the transformed, more nearly normal, data. Common transformations, in increasing order of 'strength', are square root, natural logarithm and reciprocal. This is often an entirely satisfactory approach and one that we have already used in Section 6.9, where we used the square roots of the scores from a behaviour questionnaire. Histograms for the raw behaviour scores and square-rooted scores are shown in Display 8.1. Although the square-root transformation is an improvement, no transformation will redistribute the spike of scores all at zero. Moreover, there are several situations where transforming the data may not be ideal. One is when both means and covariances are being modelled and the transformation that makes the data most nearly multivariate normal is not that which makes the model for the means linear. Another is where the original scale is 'natural' or has some practical or theoretical significance for which a model in the original metric may be preferred. In these cases, analysis on the original metric using a method of model estimation and inference that makes weaker assumptions than that of multivariate normality may be desirable.

Broadly speaking, inappropriately assuming the data to be MVN does less damage to the properties of the parameter estimates (for example, they often remain consistent) than it does to the standard methods for estimating their precision – the standard errors. One approach is, therefore, to estimate the model using, say standard maximum likelihood, but then to obtain estimates of the standard errors by a method robust to this incorrect model specification. Jackknife and bootstrap methods are two possibilities described more fully in the next section. Both can be computationally time-consuming, and perhaps more easily implemented is a 'sandwich' parameter covariance estimator, such as that of White (1982) and further developed by others (e.g. Browne and Shapiro, 1988; Satorra and Bentler, 1990). This is easily obtained from EQS simply by the addition of ',ROBUST' after the METHOD=ML specification.

Display 8.2 compares the results obtained from the model of Section 6.9 for the effects of reading on behaviour and of behaviour on reading using the raw behaviour score, the square-root transformed behaviour score, and the robust method with the raw score. All three analyses give very similar values for the

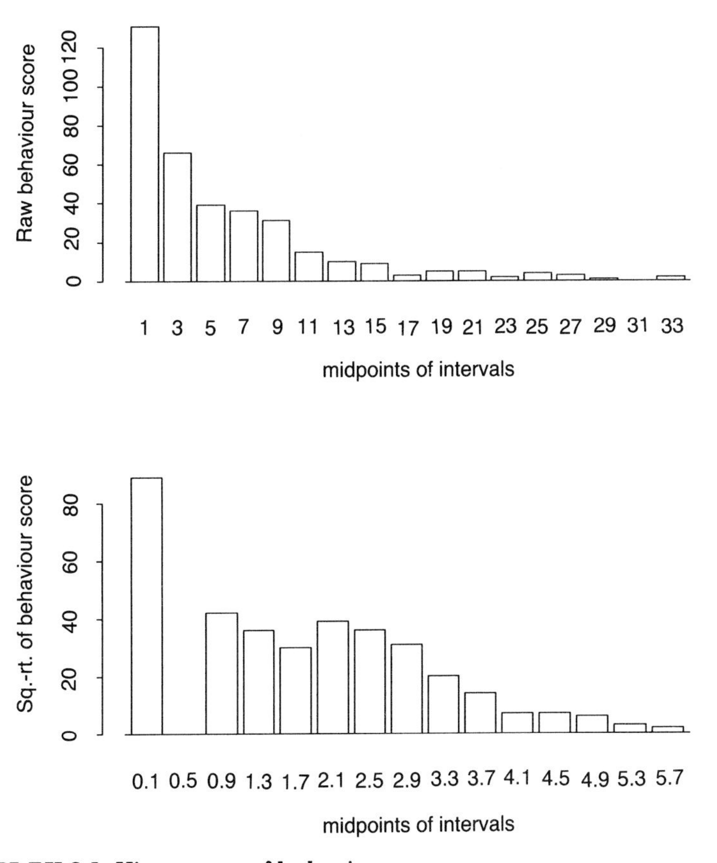

DISPLAY 8.1 Histograms of behaviour scores.

goodness of fit, indicating a fair but not excellent fit to the data. The robust method, since it is only concerned with the parameter covariance matrix, always gives the same parameter estimates as the basic estimation method with which it is used (here maximum likelihood), and here has given z-statistics of slightly larger magnitude. The transformed data give larger magnitudes for both estimated effects and z-statistics.

A second approach is to estimate the model under a more general model than one that assumes the data to be MVN. The arbitrary distribution generalized least squares (AGLS) method of estimation, in addition to using covariances, also uses information about the variables that relates to their skewness and kurtosis – that is, it takes into account some aspects of non-normality of data that are commonly found. Display 8.2 also shows the estimated effects and z-statistics for the reading–behaviour analysis using this

	Reading on Behaviour Standardized estimate	*z-test*	*Behaviour on Reading Standardized estimate*	*z-test*	*Goodness-of-fit chi-square (8 d.f.)*
ML – Raw Data	−0.214	−2.78	−0.096	−1.82	24.26
ML – Sqrt. Tran.	−0.235	−3.21	−0.113	−2.19	23.96
Robust	−0.214	−2.85	−0.096	−1.99	23.50
AGLS	−0.198	−2.36	−0.094	−1.74	18.27

DISPLAY 8.2 Alternative estimates and standard errors obtained for the reading and behaviour example.

method. This suggests that at least some of the lack of fit arises from non-normality in the data, but the estimated effects are slightly smaller and slightly less significant than those from the other methods. The evidence for the effects of behaviour on reading now looks more doubtful, but that for reading on behaviour remains strong.

It should be noted that both the procedures for robust standard errors and the AGLS method of estimation require the input data in their original case-by-case form rather than in the form of covariance matrices. In addition, the higher-order moments that the AGLS method makes use of typically have substantial sampling variances. This procedure is therefore not recommended unless the available sample is large (see Hu, Bentler and Kano, 1992 for further discussion).

8.3.2 Categorical data

The most common procedure that is adopted for dealing with categorical data is to compute some form of correlation matrix, and then to proceed as if the data had been obtained from continuous variables. We discuss some of the problems involved in each of these two steps.

The easiest correlation to compute is the simple **Pearson** correlation, that is, that used for continuous variables. This assumes the categorical codes are *point scores* on a continuous linear scale. However, if the categorical measures have arisen from the imposition of thresholds on a continuously distributed normal variable, with category values being assigned according to which *interval* between thresholds a value falls in, then the correlation wanted is the correlation with the underlying continuous variable. Where such a categorical variable is correlated with another categorical variable and with a continuous variable, the correlations are referred to as **polychoric** and **polyserial** correlation coefficients, respectively. However, where there are more than two categorical variables, the numerical computation involved in producing this matrix becomes considerable. Several programs that calculate estimates of these correlation coefficients in fact perform the calculations separately for each pair of variables. This is equivalent to allowing different thresholds to be used when the same categorical variable is correlated with one variable from those used when it is correlated with another – clearly unsatisfactory.

Display 8.3 illustrates the values obtained for the correlation coefficients from these two methods and from the original continuous measures – the raw (rather than standardized) score obtained on the three reading tests for 362 boys analysed in the previous section and in Section 6.9. The categorical scores were obtained by grouping all those below a score of 80 as category 1, those above 100 as category 3 and those between as category 2. The polychoric correlations are clearly much closer to those obtained from the continuous raw score data, than the standard Pearson correlations.

For such categorical data the sampling variances of the correlations calculated by either method will be larger than would

	Age 10–age 13	*Age 10–age 14*	*Age 13–age 14*
Pearson with raw score	0.747	0.758	0.864
Pearson with 3 categories	0.559	0.624	0.699
Polychoric with 3 categories	0.741	0.774	0.871

DISPLAY 8.3 Correlation coefficients among continuous and categorical reading test scores.

have been obtained from the same number of cases on equivalent continuous measures. Thus, to use these correlation matrices specifying them to be based on the actual sample size (using CASES=sample size in EQS) exaggerates the precision of the data, leading to inflated significance levels and underestimated standard errors. Model fitting programs can be informed of the varying level of precision in the correlations either by use of a weight matrix or by using a method of determining the precision of estimates that is more directly based on the raw data (Lee *et al.*, 1992). In EQS we might try the use of METHOD=ML,ROBUST illustrated in the previous section. However, the model we have chosen to illustrate, shown in Display 8.4, involves the fitting of a latent growth curve model to the reading test score data. This model requires the fitting of both means and covariances, and for such a task neither the robust method nor the AGLS method can

```
/TITLE
 Latent Growth Model for Reading Scores
/SPEC
 CASES=362; VARIABLES=3; MATRIX=RAW; ANALYSIS=MOM;
 METHOD=ML; DATA='filename';
/SIMULATION
 BOOTSTRAP=362; REPLICATIONS=100;
/LABELS
 V1=READ10; V2=READ13; V3=READ14;
/EQUATIONS
 V1=F1+E1;
 V2=F2+E2;
 V3=F3+E3;
 F1=F4;
 F2=F4+F5;
 F3=F4+1.333*F5;
 F4=20.0*V999+D4;
 F5=1.0*V999+D5;
/VARIANCES
 E1 to E3=10.0*;
 D4=100.0*; D5=10.0*;
/CONSTRAINTS
 (E1,E1)=(E2,E2)=(E3,E3);
/OUTPUT
 PA;
/END
```

DISPLAY 8.4 EQS set-up for latent growth curve model with bootstrap resampling.

be used. Instead, Display 8.4 illustrates the EQS commands for **bootstrap resampling**.

Bootstrap resampling involves the repeated drawing of samples, with replacement, from the cases actually observed, followed by fitting the model to each such sample. In each bootstrap sample, the sampling is done with replacement, allowing each case to be drawn once, more than once or not at all. Usually, as here, the size of the sample drawn is equal to the actual sample size (if the sampling were done without replacement, the same and complete sample would be drawn every time). In this application, the resampling is replicated 100 times and the output from each of these 100 such resamplings and model fittings is sent in very condensed form to the output file EQSOUT.DAT. The contents and format of this file are described in the regular EQS output file. Under a variety of circumstances, particularly in small samples and where the distributional assumptions of the model are not entirely met (as here), the results obtained by this resampling method have been shown to be better than those obtained from simply applying ML to the one original sample (Efron, 1981; Boomsma, 1986). In

	Goodness-of-fit				*Non-linearity*
	Chi-squared	*Normed index*	*Estimate F3,F5*	*Standard error*	*z-test (Est-1.333)/SE*
Original continuous raw scores	20.09	0.98	1.445	0.035	3.20
Pearson correlation for 3 categories	35.87	0.92	1.373	0.052	0.77
Polychoric correlation for 3 categories	53.78	0.94	1.360	0.035	0.77
Pearson correlation* for 3 categories and 100 bootstrap replicates	39.30	0.92	1.364	0.059	0.53

* Mean chi-squared and parameter estimate, standard error obtained as the standard deviation of the bootstrap estimates

DISPLAY 8.5 Latent growth curve model applied to continuous and categorical reading scores: goodness of fit and departure from non-linearity.

particular, the standard deviation of the parameter estimates obtained over the many replications is often a better estimate of the true precision than the normal standard error.

Display 8.5 compares, for several different methods, the chi-squared goodness-of-fit statistic and normed index, the estimate of the parameter describing the slope in the growth curve from age 10 to age 14, its standard error (standard deviation for the bootstrap estimate) and the value of the z-test for testing whether this parameter is significantly different from that expected for simple linear growth with age (4 years/3 years=1.333 for these data – see Section 6.9). Applied to the original continuous raw score data, the model fits well by the standards of the fit index, though the chi-squared statistic makes this look more dubious. Using the polychoric correlations, but with means and standard deviations from the raw categorical data, results in a model that apparently fits substantially worse according to the chi-squared statistic, although the fit indices were less affected. The inflated χ^2 and the fact that the standard error for the slope parameter is identical to that obtained using the continuous measures, rather than larger, are both likely to be the consequence of the model's not having been properly informed of the increased sampling variance that would be expected from categorical rather than continuous data. Results obtained using the Pearson correlations (equivalent to using the raw categorical data in EQS) give an intermediate value for χ^2 (but still worse indices) and larger standard error for the slope, a standard error that is increased still further if the bootstrap method is applied. Although the estimated slope parameters are similar in value in all cases, the test of linear growth was rejected only when using the original continuous methods. In part, this reflects the lower statistical power that categorical measures offer as compared to equivalent continuous measures. The question of power is returned to in Section 8.6.

8.4 MISSING DATA

The choice of what can be done with data sets subject to missing data depends upon how the missing data arose. Little and Rubin (1987) proposed a very useful classification, perhaps most easily understood in the context of a hypothetical example. Display 8.6 illustrates a set of three variables, measures of some trait made on

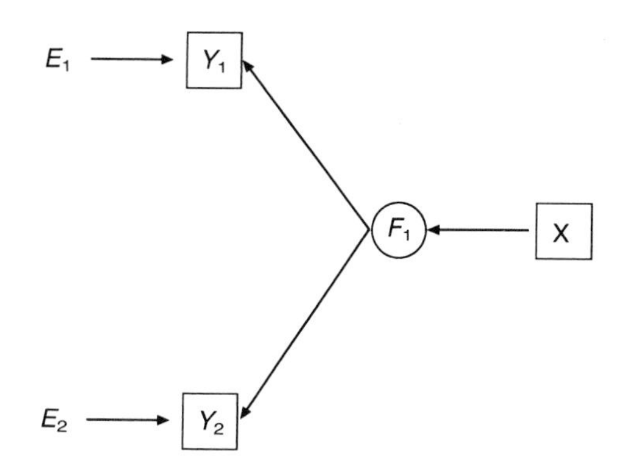

DISPLAY 8.6 A stable outcome measured on two occasions with a single causal variable.

two occasions and denoted by Y1 and Y2, and a single causal variable X. The values of the two outcome variables depend upon their shared common factor F1 and upon errors E1 and E2.

Little and Rubin consider three types of missing data, in order of increasing intractability. Data **missing completely at random** (MCAR) refers to the situation where the probability that a particular observation is missing does not depend upon either its own value (were it observed) or the value of any other observation. Thus, the probability of there being a missing value on any variable, say Y1, should not vary with any of the other variables, including the errors and factor score F1.

For data **missing at random** (MAR) the probability of an observation being missing may depend upon the values of the other observed measures but does not depend upon those unobserved. Thus, the probability of being missing on Y2 could, for example, vary with the value of Y1. However, the probability of being missing should not vary with the value of Y2, the factor score F1 or the error term E1, since these are not observed.

In the case of **non-ignorable missing** data, the probability of an observation being missing depends not only on the values of the observed variables but also on values of those unobserved.

Only certain specialized or *ad hoc* procedures are available for non-ignorable missing data. In general, methods of analysis assume missing data to be either MCAR or MAR. However, the foregoing suggests that an assumption of MAR is likely to be less wrong, sometimes substantially less wrong, than one of MCAR.

8.4.1 Dependent variables that are missing completely at random

Of course, in practice rather little may be known about the form of missing data present in any given data set. However, one situation where we do know the process generating the missing data is in sample design. For example, for reasons of economy it might be decided that only a simple random sample of the original subjects would be seen at the second occasion. All those unseen would then have missing values for Y2. However, the probability of the data being missing is constant from case to case, at 1 minus the proportion followed up, and does not depend on the values of either observed or unobserved variables. The missing data would thus be MCAR. Alternatively, the score on the first occasion might be used as a screening instrument and a sample selected to be seen at the follow-up that was stratified by the value of Y1. In the most extreme case it might be decided only to follow up those with high scores. Here again, all those not followed up will have missing values for Y2 but the probability of their being missing clearly depends only on the value of the always observed variable Y1. Thus the missing data would be MAR.

In the case of the missing data being MCAR, the approach of analysing only those cases with complete data is the simplest and usually wholly satisfactory. Occasionally, the information contained in the incomplete data cases is too great to be ignored and effective use of it can substantially improve the efficiency and power of the analysis. To illustrate how this can be done some data were simulated that conformed to the model shown in Display 8.6. We have chosen to use simulated data, rather than an empirical data set, because then the true values of the parameters that the fitted models should be attempting to recover are known. The relative advantages of different methods can thus be shown more convincingly. The data were generated using the simulation facility within EQS using the command file shown in Display 8.7. The /SIMULATION paragraph specifies a SEED or initial value for the pseudorandom number generator, that we want a single sample or REPLICATE to be drawn (the CASES=1000 statement specifies the size of sample to be drawn), that the sample should be drawn from a POPULATION that conforms to the MODEL specified in the rest of the command file, and that the DATA generated should be written to a file beginning with the letters 'SIM' – in fact the file SIM001.DAT for this one and only replicate.

```
/TITLE
 Simulating Data in EQS for the Model of Display 8.6
/SPECIFICATIONS
 CASES=1000; VARIABLES=3;
/LABELS
 V1=Y1; V2=Y2; V3=X;
/SIMULATION
 SEED=341215;
 REPLICATIONS=1;
 DATA='SIM';
 POPULATION=MODEL;
/EQUATIONS
 V1=F2+E1;
 V2=F2+E2;
 F2=F1+D2;
 V3=F1;
/VARIANCES
 D2=1.0; F1=1.0;
 E1=2.0*; E2=2.0*;
/END
```

DISPLAY 8.7 EQS command file to generate data consistent with the model of Display 8.6.

The data actually generated are the values of V1, V2 and V3 for 1000 cases.

To begin with, we shall focus solely on the two variables Y1 and Y2, and consider the estimation of a simple one-factor model. The model of Display 8.8, fitted to all 1000 cases, gave the first column of results shown in Display 8.9. Making 950 randomly selected cases 'missing' on the value of Y2 left just 50 complete data cases, which when analysed gave the results also shown in Display 8.9. To include the other 950 cases with values for Y1 but not Y2, they were formed into a second group and analysed using the multi-groups option as shown in Display 8.10.

There are two main features to notice. First, the equations and variance paragraphs (and covariance paragraph, were any estimated) for the second group include only those variables and parameters actually required for the model of the reduced set of observed variables. Specifically, there is no equation for V2, and no variance for E2. Second, there are cross-group constraints (as employed in previous multigroup analyses) and in this case every

```
/TITLE
 Complete Data Example
/SPECIFICATIONS
 CASES=1000; VARIABLES=2; MATRIX=CORR; ANALYSIS=COV;
/LABELS
 V1=Y1; V2=Y2;
/EQUATIONS
 V1=F1+E1;
 V2=F1+E2;
/VARIANCES
 E1 TO E2=1.0*; F1=1.0*;
/MEANS
 0.066 0.009
/STANDARD DEVIATIONS
 2.08 1.96
/MATRIX
 1.000
 0.517 1.000
/END
```

DISPLAY 8.8 Simple measurement model for complete data on two measures.

	Population values	*Whole sample analysis (N = 1000)*	*Complete data cases only (N = 50)*	*MCAR multigroup analysis (N = 50 and 950)*
F1 Variance	2.00	2.11 (0.15)	1.97 (0.56)	2.10 (0.42)
E1 Variance	2.00	2.22 (0.15)	2.07 (0.57)	2.22 (0.42)
E2 Variance	2.00	1.73 (0.14)	0.85 (0.43)	0.79 (0.38)
Chi-squared		0	0	Unknown
d.f.		0	0	1

DISPLAY 8.9 Results for the analysis of MCAR data.

parameter involved in the model for the incomplete data group must be set equal to its corresponding parameter in the complete data group and must be set to share the same starting values. The results of the analysis with both complete and incomplete data are also shown in Display 8.9. The inclusion of the incomplete data

```
/TITLE
 MCAR Multigroup Example – Complete Data Group
/SPECIFICATIONS
 CASES=50; VARIABLES=2; MATRIX=CORR; ANALYSIS=COV;
 GROUPS=2;
/LABELS
 V1=Y1; V2=Y2;
/EQUATIONS
 V1=F1+E1;
 V2=F1+E2;
/VARIANCES
 E1=2.0*; E2=2.0*; F1=1.0*;
/PRINT
 EFFECT=YES;
/MEANS
 .0704 −0.0923
/STANDARD DEVIATIONS
 2.0149 1.6794
/MATRIX
 1.000
 0.5829 1.000
/END
/TITLE
 MCAR Multigroup Example – Missing or Incomplete Data Group
/SPECIFICATIONS
 CASES=950; VARIABLES=1; MATRIX=CORR; ANALYSIS=COV;
/LABELS
 V1=Y1; V2=Y2;
/EQUATIONS
 V1=F1+E1;
/VARIANCES
 E1=2.0*; F1=1.0*;
/CONSTRAINTS
 (1,E1,E1)=(2,E1,E1);
 (1,F1,F1)=(2,F1,F1);
/MEANS
 0.07
/STANDARD DEVIATIONS
 2.08
/MATRIX
 1.0
/END
```

DISPLAY 8.10 Simple measurement model for data missing completely at random on one measure.

```
/TITLE
 MAR Multigroup Example – Complete data group
/SPECIFICATIONS
 CASES=50; VARIABLES=2; MATRIX=CORR;
 ANALYSIS=MOMENT; GROUPS=2;
/LABELS
 V1=Y1; V2=Y2;
/EQUATIONS
 V1=F1+E1;
 V2=F1+E2;
 F1=0.0*V999+D1;
/VARIANCES
 E1=2.0*; E2=2.0*; D1=1.0*;
/PRINT
 EFFECT=YES;
/MEANS
 .0704 -0.0923
/STANDARD DEVIATIONS
 2.0149 1.6794
/MATRIX
 1.000
 0.5829  1.000
/END
/TITLE
 MAR Multigroup Analysis – Missing or Incomplete Data Group
/SPECIFICATIONS
 CASES=950; VARIABLES= 1; MATRIX=CORR; ANAL=MOMENT;
/LABELS
 V1=Y1; V2=Y2;
/EQUATIONS
 V1=F1+E1;
 F1=0.0*V999+D1;
/VARIANCES
 E1=2.0*; D1=1.0*;
/CONSTRAINTS
 (1,E1,E1)=(2,E1,E1);
 (1,D1,D1)=(2,D1,D1);
 (1,F1,V999)=(2,F1,V999);
/MEANS
 0.07
/STANDARD DEVIATIONS
 2.08
/MATRIX
 1.0
/END
```

DISPLAY 8.11 Simple measurement model for data missing at random on one variable.

cases has improved the precision of the parameter estimates, as indicated by the smaller standard errors. However, even with the very substantial additional incomplete sample the improvements may not seem worth the extra model-fitting effort.

8.4.2 Dependent variables missing at random (MAR)

The previous analysis assumed the data to be MCAR. Extending the model as shown in Display 8.11 to analyse the means as well as the covariances, again constraining *all* parameters to be equal across groups, allows the data to be analysed under the assumption that the missing data are MAR rather than MCAR. If the fitted model is the correct one, then the chi-squared goodness-of-fit statistic for this model is a test of whether the missing data are in fact MCAR.

Fitting the model of Display 8.11 to the current simulated data gave a chi-square of only 0.76 with 3 d.f., confirming what we knew already, that the missing data were MCAR.

```
Screened positive complete data cases (N = 524)

/MEANS
 1.6382 0.7922 0.3859
/STANDARD DEVIATIONS
 1.2570 1.7914 0.9203
/MATRIX
 1.000
 0.3027 1.000
 0.3325 0.4214 1.000
/END

Screened negative incomplete data cases (N = 476)

/MEANS
 -1.66
/STANDARD DEVIATIONS
 1.28
/MATRIX
 1.0
/END
```

DISPLAY 8.12 Data obtained when only those with Y1 above zero were followed up.

	Population values	Whole sample analysis (N = 1000)	MAR multigroup analysis (N = 524 and 476)	Complete data cases only (N = 524)
F1 Mean	0	0.03 (0.06)	0.08 (0.06)	1.42 (0.05)
F1 Variance	2	2.11 (0.15)	1.94 (0.18)	0.54 (0.11)
E1 Variance	2	2.22 (0.15)	2.39 (0.19)	1.09 (0.12)
E2 Variance	2	1.74 (0.14)	1.85 (0.19)	3.05 (0.22)
Chi-squared		0.08	986.95	99.22
d.f.		1	3	1

DISPLAY 8.13 Results for the analyses involving missing data.

Display 8.12 gives similar data to those just analysed but now the follow-up measure, Y2, was only obtained for those with a score on Y1 above zero, roughly the top 50% of subjects. Thus those with scores below zero all have missing values for Y2. Not surprisingly, the mean scores on Y1 are now very different between the two groups, and to a lesser degree so is the standard deviation. However, since the probability of being missing depends only on the observed value of Y1, the data are still MAR. Fitting the model of Display 8.11, but now with the data of Display 8.12, recovers the correct parameter estimates as shown by the results given in Display 8.13. The remarkable success of this procedure is made all the more persuasive by the hopeless results also shown in Display 8.13, obtained from fitting a single-group model to the complete data cases only. Unfortunately, although differences in model chi-squared statistics are still useful, the overall model χ^2 for the two-group approach with MAR data cannot be expected to be small, even if the model is correct, since its value reflects the degree to which the missing data are not MCAR. For the simulated MAR data a value of 987.28 with 4 d.f. was obtained, thus confirming the appropriateness of a particular model when data are MAR (rather than MCAR) is made more difficult.

8.4.3 Missing independent, explanatory or causal variables

In the previous examples the only equations that were required for the incomplete data group were those involving variables that were

```
/TITLE
 MAR Multigroup Analysis – Screened Positive Complete Data
 Group
/SPECIFICATIONS
 CASES=524; VARIABLES=3; MATRIX=CORR; ANAL=MOMENT;
 GROUPS=2;
/LABELS
 V1=Y1; V2=Y2; V3=X;
/EQUATIONS
 V1=F1+E1;
 V2=F1+E2;
 F1=0.0*V999+1.0*F2+D1;
 F2=0.0*V999+D2;
 V3=F2;
/VARIANCES
 E1=2.0*; E2=2.0*; D1 TO D2=1.0*;
/PRINT
 EFFECT=YES;
/MEANS
 1.6382 0.7922 0.3859
/STANDARD DEVIATIONS
 1.2570 1.7914 0.9203
/MATRIX
 1.000
 0.3027 1.000
 0.3325 0.4214 1.000
/END
/TITLE
 MAR Multigroup Analysis – Screened Negative Incomplete Data
 Group
/SPECIFICATIONS
 CASES=476; VARIABLES=1; MATRIX=CORR; ANAL=MOMENT;
/LABELS
 V1=Y1; V2=Y2; V3=X;
/EQUATIONS
 V1=F1+E1;
 F1=0.0*V999+1.0*F2+D1;
 F2=0.0*V999+D2;
/VARIANCES
 E1=2.0*; D1 TO D2=1.0*;
/CONSTRAINTS
 (1,E1,E1)=(2,E1,E1);
 (1,D1,D1)=(2,D1,D1);
 (1,D2,D2)=(2,D2,D2);
 (1,F1,V999)=(2,F1,V999);
 (1,F2,V999)=(2,F2,V999);
 (1,F1,F2)=(2,F1,F2);
/MEANS
 -1.66
/STANDARD DEVIATIONS
 1.28
/MATRIX
 1.0
/END
```

DISPLAY 8.14 Model for data missing at random on both dependent and independent measures.

not missing, with Y1 being predicted only by the common factor F1 and an error term. But what if among the predictors of such a variable there is another variable that is missing? For example, consider the situation where it was only at the follow-up that the causal variable X, thought to be related to F1, was measured. In the complete data group the factor F1 will now be a dependent variable in an equation in which X will be a predictor. The variance of F1 is no longer a parameter that can be constrained across groups, but has a mean and variance that are predicted by a variable that has no corresponding variable in the incomplete data group. The solution to this is similar to that introduced in Section 6.7, for repeated measures of ability that were unequally spaced in time at ages 6, 7, 9 and 11. In that example phantom latent variables were introduced that themselves had no direct observed or manifest variable. The parameters that defined the phantom factors, their variances and the coefficients that linked them to other variables, were identified by being constrained to be equal to other equivalent parameters for which there *were* observed variables. That this approach might be relevant here is clear from a recognition that the repeated measures example can be conceptualized as one in which the data at ages 8 and 10 were missing.

Display 8.14 shows the appropriate EQS set-up that uses this approach to solve this problem. The essential idea is to ensure that

	Population values	*MAR multigroup analysis (N = 524 and 476)*	*Complete data cases only (N = 524)*
F1 Mean	0	0.07	1.37
F2 Mean	0	0.01 (0.04)	0.39 (0.04)
D1 Variance	1	0.91 (0.14)	0.19 (0.09)
D2 Variance	1	1.00 (0.06)	0.85 (0.05)
E1 Variance	2	2.35 (0.18)	1.30 (0.12)
E2 Variance	2	1.89 (0.17)	2.84 (0.20)
F1–F2 Covariance	1	1.01 (0.06)	0.57 (0.05)
Chi-squared		987.28	116.85
d.f.		4	4

DISPLAY 8.15 Results for the causal model with risk factor only for screen positive subjects: parameter estimates and standard errors.

all the equations that need to appear for the incomplete data group contain only variables that are not missing, or factors that are predicted by other equations that conform to these rules, or factors whose mean and variance are parameters that can be directly constrained equal to their values in the complete data group. In practice, this involves the addition of an equation for each causal variable that is missing in the incomplete data group that makes the observed variable identical with that factor. This factor, not the missing manifest variable, then appears in the equations for the incomplete data group.

Display 8.15 shows the results of fitting this model and also those obtained from analysing only the complete data cases. The superiority of the multigroup method and the corresponding hopeless inadequacies of the complete data case approach are obvious.

8.5 THE IMPLICATIONS OF SAMPLING DESIGN: WHICH DATA CAN AND CANNOT BE EASILY ANALYSED?

No mention has been made as to the relationship between the form of the model being estimated and the sampling design used to obtain the data. This is an important omission, because most programs, as they currently stand, are able to estimate models appropriate to data obtained under only a few very simple sampling designs. We discuss this in relation to some common designs, all of which use probability sampling.

1. *Simple random sampling of one or more populations*. This has been the design implicitly assumed in every one of the analyses of this book. Where two or more groups have been used, the data in each group have been assumed to be a random sample drawn from a distinct population where the variables under analysis are assumed to be multivariate normally distributed.
2. *Random sampling of a population stratified on the value of a continuous variable*. Such samples are extremely common, often arising, for example, in epidemiology as the result of oversampling from those who score positive on some screening instrument, questionnaire or interview. The popular approaches are either just to analyse the screened positive sample or to attempt to compute a single set of summary statistics (such as the covariance matrix) using weights. Neither of these is

entirely satisfactory – for a critique of weighting in longitudinal data analysis, see Fienberg (1980) and Hoem (1985). But as we have seen in Section 8.4, an alternative approach is provided by the multigroup procedure, which includes those examined with just the screening instrument and whatever other variables were measured in one group, and those who screened positive and examined more intensively in another group.

3. *Random sampling of a population stratified on a truly categorical variable.* Again the most common approach is to analyse a single covariance matrix obtained by reweighting the data from each stratum in proportion to the reciprocal of the sampling fraction. There seems to be no logical basis for doing this if the researcher is interested in process. Consider a sample stratified into boys and girls. If the process is the same for boys as for girls, then weighting each stratum differently is pointless – essentially the same answer is obtained if the boys alone were analysed by giving zero weight for girls, as when the girls alone are analysed by giving zero weight to the boys, and by using any intermediate weighting. If the process for boys is different from that for girls, then we ought to want to know about these differences – the process for some hypothetical child of intermediate sex, identified by the analysis of weighted data, may be quite meaningless. Whether the process is the same or not, the analysis should proceed using the multigroup procedure with unweighted data. The procedures illustrated in Chapter 7, in which parameters are constrained to be equal across the groups, can then be used to determine in what respects the strata differ. If effects (say, of some variable that might be manipulated by policy) are found to differ by stratum, then the average effect of some proposed policy change can be found by a suitable weighted average of the estimated effects within each stratum. In other words, weights may be relevant to the interpretation of models but should not be used in model estimation.
4. *Dependent samples, e.g. pairwise case-control matching.* In general these cannot be analysed using the two-group approach. For simple problems the model can be specified using pairs of variables, one for cases and one for controls, and by allowing each pair of variables to be correlated. Display 8.16 illustrates a path diagram for a model allowing a test of the difference in the mean of a latent variable within two matched samples. Variables V1 and V2 are the indicators for one group and V3 and V4 are the indicators for the other. Since the samples have

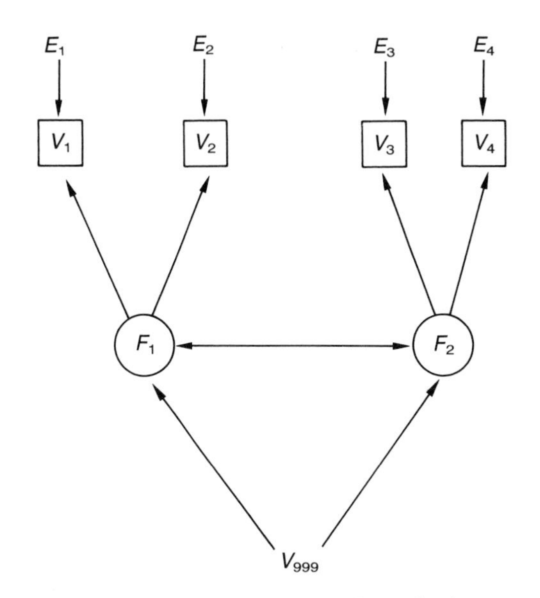

DISPLAY 8.16 Path diagram for a model to obtain a test of difference of means on a latent variable with two indicators: test for matched-pairs data.

been matched, the factor scores are likely to be correlated (as shown) and, in addition, measurement errors (say where V1 and V3 are the same instrument and V2 and V4 likewise) might also be correlated (not shown). A test of difference in mean is obtained by comparing the fit of a model with F1,V999 constrained equal to F2,V999 to that obtained without the constraint.

5. *Dependent samples, e.g. clustering.* Many samples are obtained as the result of some multistage sampling procedure (for example, children drawn from a sample of schools). There is no wholly satisfactory way of analysing such data for models of any realistic complexity using current programs of the EQS type. The use of the ROBUST procedure, described in Section 8.3.1, usually affords some protection against the inflated significance levels and exaggerated levels of precision that ignoring the design usually causes.

8.6 STATISTICAL POWER

The social and behavioural research literature is now full of research papers fitting increasingly complicated models. With such complicated models it is very hard to gain any impression of whether the authors had any realistic chance of rejecting their null hypothesis or discriminating between one model and another. However, there are several ways in which this may be explored.

The example we consider, similar to that of Neale and Cardon (1992), is the ability to detect a genetic component that explains the similarity of values found in identical and non-identical twins on some measure. The total variance in the observed values for a twin is made up of three components: that due to genes (A); that due to environmental factors shared by both members of a twin pair (C); and that due to environmental factors that are not shared (E). The latter component includes measurement error. Non-identical twins only share half their genes while identical twins have all their genes in common. Thus, under a simple model of additive components of variance:

Total variance	$= A + C + E$
Covariance for identical twins	$= A + C$
Covariance for non-identical twins	$= 0.5A + C$

We might guess that 40% of the variance of some measure was due to genetic variance, 20% due to common environment and 40% due to other unshared factors. If the measure was standardized to have variance 1 then the covariance matrices expected under the model would be:

Identical twins		Non-identical twins	
1.0		1.0	
0.6	1.0	0.4	1.0
(0.4+0.2)		(0.2+0.2)	

Display 8.17 shows the path diagram as it is typically drawn in behavioural genetic analyses, and the EQS equations for specifying the corresponding model are shown in Display 8.18. Essentially, the equations for the predicted scores of each twin are all the same, but in the case of the identical twins the genetic effects are forced to be completely correlated (that is, the same genes) while in the non-identical twins the genetic effects are forced to have a covariance of 0.5 (that is, they share half their genes). The shared environment effect appears as the same factor throughout.

To show that the measure has a genetic component at all, we have to show that a model in which the variance of A is reduced to zero fits significantly worse than one in which it is left as a free

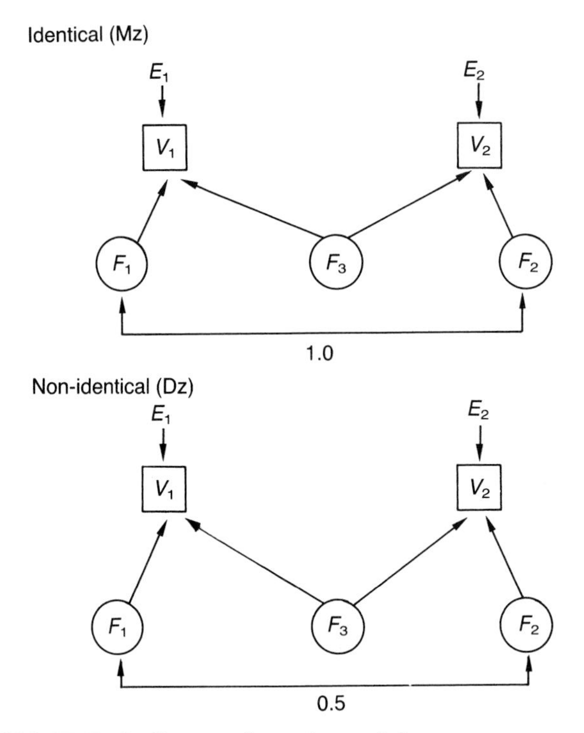

DISPLAY 8.17 Path diagram for twin model.

parameter. With a very small number of twins this would be hard to do. How many pairs of twins are required in order for there to be an 80% chance of detecting the genetic component as significantly different from zero (known as 80% power) at the 5% level of significance (and assuming, for simplicity, equal numbers of identical and non-identical twins)? Alternatively, what is the power of a particular sample size (say, 100 pairs of identical twins and 100 non-identical) for detecting the genetic effects as significant at the 5% level?

Specifying CASES=100 and fitting the model of Display 8.18 gives a chi-squared value of 0 for the unrestricted model and of 4.48 for the model with genetic effects removed (for example, by setting all the coefficients on F1 and F2 to zero).

However, the input data we have used are those expected under the true model (incorporating 40% genetic variance) without any allowance for fluctuations from those expected values due to sampling. Thus this is not the usual chi-squared test statistic but an

```
/TITLE
 GENETIC TWIN ANALYSIS: GROUP1=identical GROUP2=non-
 identical
/SPECIFICATIONS
 GROUPS=2; CASES=100; VARIABLES=2; METHOD=ML;
/LABELS
 F1=TWIN1GENES; F2=TWIN2GENES; F3=COMMONENV;
/EQUATIONS
 V1=0.4*F1+0.1*F3+E1;
 V2=0.4*F2+0.1*F3+E2;
/VARIANCES
 F1=1.0; F2=1.0; F3=1.0;
/COVARIANCES
 F1,F2=1.0;
/CONSTRAINTS
 (V1,F1)=(V2,F2);
 (V1,F3)=(V2,F3);
 (E1,E1)=(E2,E2);
/MATRIX
 1.0
 0.6 1.0
/END
/SPECIFICATIONS
 CASES=100; VARIABLES=2; METHOD=ML;
/LABELS
 F1=TWIN1GENES; F2=TWIN2GENES; F3=COMMONENV;
/EQUATIONS
 V1=0.4*F1+0.1*F3+E1;
 V2=0.4*F2+0.1*F3+E2;
/VARIANCES
 F1=1.0; F2=1.0; F3=1.0;
/COVARIANCES
 F1,F2=0.5;
/CONSTRAINTS
 (V1,F1)=(V2,F2);
 (V1,F3)=(V2,F3);
 (E1,E1)=(E2,E2);
 (1,V1,F1)=(2,V1,F1);
 (1,V1,F3)=(2,V2,F3);
 (1,E1,E1)=(2,E2,E2);
/MATRIX
 1.0
 0.4 1.0
/END
```

DISPLAY 8.18 EQS model for estimating genetic, shared environmental and specific effects on twins.

How cow documentaries are made

The FAR SIDE

July

30

SUNDAY

	Power to detect difference at 5% significance level				
	0.50	*0.60*	*0.70*	*0.80*	*0.90*
1 d.f. test	3.84	4.90	6.17	7.85	10.51
2 d.f. test	4.96	6.21	7.70	9.64	12.65

DISPLAY 8.19 Values of the non-centrality parameter from Pearson and Hartley (1972).

expected' χ^2, a quantity that, when used in conjunction with tables of the non-central chi-squared distribution (see, for example, Pearson and Hartley, 1972), can be used to answer questions about power. In that context the expected χ^2 is referred to as the **non-centrality parameter**. Display 8.19 gives values from the Pearson and Hartley tables of this non-centrality parameter required to give certain levels of power for 1 d.f. and 2 d.f. tests at the 5% significance level. With samples of 100 of each kind of twin pair, interpolation within Display 8.19 suggests that we would have only a 57% chance of detecting the genetic effects.

The display also indicates that we would have required a non-centrality parameter of 7.85 from the previous analysis to have attained 80% power. We could obtain such a value by increasing the sample size. Now the χ^2 obtained from a test generally increases in proportion to the sample size, so that we can obtain an estimate of the sample size required by the formula:

Reequired sample size = The product of the non-centrality parameter for 80% power and the sample size used for the expected χ^2, divided by the expected χ^2

In our case, this suggests that $7.85 \times 100/4.48 = 175$ pairs of each sort of twin would be required for there to be an 80% chance of detecting the genetic effects as significant at the 5% level. Studies much smaller than this may clearly be failing to find significant evidence for genetic effects even when, as here, the measure being analysed is under a substantial degree of genetic control.

8.7 SUMMARY

This chapter has given some simple practical hints and, all too briefly, has covered a number of more advanced issues relating to non-normal data, missing data and sample design, and statistical power. Research on the performance of current methods and on potentially better methods of dealing with non-normal data is continuing apace. Our treatment of this topic should be considered more as an introduction to the problem rather than any sort of final word. The availability of more general treatments of missing data is also a likely prospect, in particular of missing values scattered across the data matrix rather than conforming to the simple patterns examined here. Our examples suggest that methods for accounting for missing data when they are missing at random (rather than simply missing completely at random) are a powerful tool. However, it should be noted that this may be achieved only at the cost of a greater dependence on other model assumptions, notably multivariate normality. Finally, the discussion of statistical power extends the relevance of the structural equation modelling approach to practical issues of study design.

Though this book began by assuming relatively little knowledge of multivariate statistics, it has progressed to a level where quite complicated structural equation models have been fitted. The style of this book has been deliberately 'enabling', intended to allow readers actually to fit models and perform analyses. In the process of gaining this practical skill and familiarity with latent variable modelling concepts, we hope that readers have also taken note of our indications as to where pitfalls and problems lie. In particular we would not have wanted to encourage either 'blind model fitting' or the 'one-tool technician'. Rather, the fitting of structural equation models should occur within a broader context of a variety of exploratory and modelling methods and should draw upon wider experience and knowledge of multivariate statistics. None the less, we hope that the examples have proved sufficiently inspiring for readers to want to apply these and similar models to their own data (see special issues of Child Development, 1987 and Behavior Genetics, 1989 for other examples). This chapter has been primarily aimed at such readers, to ease them into the real world of more complicated and messy data. We wish them success!

References

Bentler, P.M. (1989) *EQS: Structural Equations Manual*. BMDP Statistical Software, Inc. Los Angeles.

Boomsma, A. (1986) On the use of bootstrap and jackknife in covariance structure analysis. *Compstat 86*. pp. 205–10. Physica-Verlag, Heidelberg.

Browne, M.W. and Shapiro, A. (1988) Robustness of normal theory methods in the analysis of linear latent variable models. *British Journal of Mathematical and Statistical Psychology*, **41**, 193–208.

Campbell, D.T. and Fiske, D.W. (1959) Convergent and discriminant validation by the multitrait-multimethod matrix. *Psychological Bulletin*, **56**, 81–105.

Cudeck, R. (1989) Analysis of correlation matrices using covariance structure models. *Psychological Bulletin*, **105**, 317–27.

Dunn, G. (1989) *Design and Analysis of Reliability Studies: The Statistical Evaluation of Measurement Errors*. Edward Arnold, London.

Dunn, G. (1992) Design and analysis of reliability studies. *Statistical Methods in Medical Research*, **1**, 123–57.

Efron, B. (1981) Nonparametric estimates of standard error: the jackknife, the bootstrap and other methods. *Biometrika*, **68**, 589–99.

Everitt, B.S. and Dunn, G. (1991) *Applied Multivariate Data Analysis*. Edward Arnold, London.

Fienberg, S.E. (1980) The measurement of crime victimization: prospects for panel analysis of a panel survey. *The Statistician*, **29**, 313–50.

Goldberg, D.P. (1972) *The Detection of Psychiatric Illness by Questionnaire*. Oxford University Press, London.

Hoem, J.M. (1985) Weighting, misclassification, and other issues in the analysis of survey samples of life histories. In *Longitudinal Analysis of Labour Market Data*. (Eds. J.J. Heckman and B. Singer). Cambridge University Press, Cambridge.

Hu, L., Bentler, P.M. and Kano, Y. (1992) Can test statistics in covariance structure analysis be trusted. *Quantitative Methods in Psychology*, **112**, 351–62.

Huba, G.J., Wingard, J.A. and Bentler, P.M. (1981) A comparison of two latent variable causal models for adolescent drug use. *Journal of Personality and Social Psychology*, **40**, 180–93.

Judd, C.M. and Milburn, M.A. (1980) The structure of attitude systems in the general public: comparisons of a structural equation model. *American Sociological Review*, **45**, 627–43.

Kluegel, J.R., Singleton, R. and Starnes, C.E. (1977) Subjective class identifications: a multiple indicator approach. *American Sociological Review*, **42**, 599–611.

Lee, S.Y., Poon, W.-Y. and Bentler, P.M. (1992) Structural equation models with continuous and polytomous variables. *Psychometrika*, **57**, 89–105.

Little, R.J.A. and Rubin, D.B. (1987) *Statistical Analysis with Missing Data*. John Wiley and Sons, New York.

Loehlin, J.C. (1987) *Latent Variable Models: An Introduction to Factor, Path, and Structural Analysis*. Lawrence Erlbaum Associates, Hillsdale, NJ.

Loehlin, J.C. and Vandenberg, S.G. (1968) Genetic and environmental components in the covariation of cognitive abilities. An additive model. In *Progress in Human Behavior Genetics*. (Ed. S.G. Vandenberg) pp. 261–78. John Hopkins University Press, Baltimore, MD.

McArdle, J.J. and Aber, M.S. (1990) Patterns of change with latent variable structural equation models. In *Statistical Methods in Longitudinal Research*, Vol. II. (Ed. A. van Eye) pp. 151–224. Academic Press, Boston.

Neale, M.C. and Cardon, L.R. (1992) *Methodology for Genetic Studies of Twins and Families*. Kluwer, Dordrecht.

Osbourne, R.T. and Suddick, D.E. (1972) A longitudinal investigation of the intellectual differentiation hypothesis. *Journal of Genetic Psychology*, **110**, 83–9.

Pearson, E.S. and Hartley, H.O. (1972) *Biometrika Tables for Statisticians*, Vol. 2. Cambridge University Press, Cambridge.

Satorra, A. and Bentler, P.M. (1990) Model conditions for asymptotic robustness in the analysis of linear relations. *Computational Statistics and Data Analysis*, **10**, 235–49.

Smith, D.A. and Patterson, E.B. (1984) Applications and generalization of MIMIC models to criminology research. *Journal of Research in Crime and Delinquency*, **21**, 333–52.

Sullivan, J.L. and Feldman, S. (1979) *Multiple Indicators: An Introduction*. Sage, Beverly Hills, CA.

White, H. (1982) Maximum likelihood estimation of mis-specified models. *Econometrics*, **50**, 1–25.

Yule, W., Berger, M., Butler, S., Newham, V. and Tizard, J. (1969) The WPPSI: an empirical evaluation with a British sample. *British Journal of Educational Psychology*, **39**, 1–13.

Index

EQS: Quick Reference

This card gives brief information about EQS commands. More detail is given in the EQS manual.

1./TITLE (optional)

For example:

```
/TITLE
Confirmatory factory analysis
Initial run
```

2./SPECIFICATIONS

This paragraph provides information on the number of cases, the number of input variables, and the method (or methods) of estimation desired, as well as a variety of other information to guide the EQS run.

CASES=	- number of observations, e.g. CASES=250;
VARIABLES=	- number of variables, e.g. VAR=10;
METHOD=	- selects estimation method. If no method is specified then maximum likelihood (ML) is used.

Options are as follows:

ME=ML;	- maximum likelihood
ME=LS;	- least squares
ME=GLS;	- generalized least squares
ME=ELS;	- elliptical least squares
ME=EGLS;	- elliptical GLS
ME=ERLS;	- elliptical reweighted LS
ME=any of the above, ROBUST;	- robust standard errors and chi-square
ME=AGLS;	- arbitrary distribution GLS (requires raw data)
MATRIX=	- specifies types of matrix to be input, a covariance matrix is the default.

Options are as follows:

MA=COV; - covariance matrix
MA=CORR; - correlation matrix
MA=RAW; - raw data
ANALYSIS= - specifies types of matrix to be analysed. The default is the covariance matrix.

Options are as follows:

ANAL=COV; - covariance matrix
ANAL=CORR; - correlation matrix
ANAL=MOM; - means and covariances

GROUPS= - number of groups in a multisample analysis e.g. GROUPS=3;

3./LABELS (optional)

Signals that identifying labels for the observed and/or the latent variables are to be used.
An example is:

```
/LABELS
V1=Age; V2=IQ; F1=Performance;
```

4./EQUATIONS

This section specifies the model that is to be fitted. One and only one equation is required for each dependent variable. The dependent variables may be either observed or latent, and parameters within the equations may be specified as either fixed or free (indicated*). An example of a set of equations is as follows:

```
/EQUATIONS
V1=1*F1+E1;
V2=1*F1+E2;
V3=1*F1+E3;
V4=1*F2+E4;
V5=1*F2+E5;
F1=1*F2+D1;
```

5./VARIANCES

Specifies fixed and free values for the variances of the independent variables. (Dependent variables are not allowed to have variances as parameters, whether fixed or free.) An example is as follows:

```
/VARIANCES
F1 to F3=1.0;     - variances fixed at 1
E1 to E5=0.5*;
```

6./COVARIANCES

Specifies the fixed and free covariances among independent variables. If a variable is involved in a covariance, its variance must also be specified in the /VARIANCE section. *Dependent variables cannot have covariances.* An example is as follows:

```
/COVARIANCES
F2, F1=0.3*;
E1 to E3=0.5*;    - sets E1, E3=0.5*, E1, E2=0.5* and E2, E3=0.5*
```

7./CONSTRAINTS (optional)

Specifies equalities, etc., between particular parameters. Some examples are:

```
/CONSTRAINTS
(E1,E1)=(E2,E2);          - note the brackets
(V1,F1)=(V1,F2);
(1,V1,F1)=(2,V1,F1);      - an across-group constraint
```

8./MATRIX (optional)

This paragraph is used for input of covariance or correlation matrix. Matrices are usually given in lower-triangular free format form. An example is:

```
/MATRIX
1.00
0.51   1.00
0.32   0.54   1.00
```

9./STANDARD DEVIATIONS (optional)

Reads in standard deviations. Used to convert input correlation matrix to a covariance matrix. An example is:

```
/STANDARD DEVIATIONS
1.34 1.27 3.25 1.89
```

10./MEANS (optional)

Used only when structured means models are being fitted. An example is:

```
/MEANS
5.32 4.76 5.25 5.40
```

11./LMTEST (optional)

Used to test hypotheses on the statistical necessity of restrictions that exist in a model.

12./WTEST (optional)

Used to test whether sets of parameters which are treated as free in a current model could be simultaneously set to zero without substantial loss in fit.

13./TECHNICAL (optional)

Used to alter default values for number of iterations, convergence criterion and tolerance. The options are as follows:

ITR= - maximum number of iterations (default is 30) e.g. ITR=50;
CON= - convergence criterion (default is 0.001), e.g. CON=0.01;
TOL= - tolerance to control accuracy in solving linear equations (default is 0.000001), e.g. TOL=0.001;

14./PRINT (optional)

Controls a variety of printed information that can help to make sense of a model and the quality of the estimates. Options are as follows:

/PRINT
EFFECT=YES; - prints direct and indirect effects as well as total effects
COVARIANCE=YES; - prints reproduced covariance matrix
CORRELATION=YES; - prints reproduced correlation matrix
PARAMETER=YES; - prints correlations of parameter estimates

15./END

Finishes the EQS run.